AF574514

PRACTICAL ATLAS FOR
Bacterial Identification

PRACTICAL ATLAS FOR Bacterial Identification

D. Roy Cullimore

Library of Congress Cataloging-in-Publication Data

Cullimore, D. Roy.
Practical atlas for bacterial identification / D. Roy Cullimore.
p. cm.
Includes bibliographical references and index.
ISBN 1-56670-392-1 (alk. paper)
1. Bacteria—Identification—Atlases. I. Title.
QR54 .C85 2000
579.3′022′2—dc21 00-039104
CIP

Direct all inquiries to CRC Press LLC, 2000 N.W. Corporate Blvd., Boca Raton, Florida 33431.

Lewis Publishers is an imprint of CRC Press LLC

International Standard Book Number 1-56670-392-1
Library of Congress Card Number 00-039104
Printed in the United States of America 3 4 5 6 7 8 9 0
Printed on acid-free paper

Foreword

One of the major challenges in my life as an applied microbial ecologist has been to grapple with the identification of bacteria. If there is going to become an understanding of the role of bacteria within and upon this planet, there is a fundamental need to appreciate the characteristics of bacteria strains belonging to described species. The current concept, as with so many other aspects of science, is to reduce the mode of analysis to the molecular chemical level and this would then be more precise. Since bacteria normally function within community groups (sometimes called consortia), this level of molecular "precision" may be inappropriate to the needs to comprehend the dynamics involved with the various bacteria within the site being examined.

This book gives a more generalized traditional overview of some of the practices that have become established over the years to identify bacteria. Why would this be a possible need at this time? Well, there is an inevitable need to separate and concentrate on the aggressive strains within a bacterially infested site rather than the passive. There is also a need to differentiate the potentially pathogenic from the relatively harmless saprophytes. Like a farmer tends to a crop, so often in these times there is a need to manage the many microbial "crops" that occur virtually everywhere. A better understanding of the bacteria genera involved in a particular "crop" and the nature with which they respond to management practices allows a more matured management of the bacterial events. Hence, this book is not designed to be a scholarly guide concentrating on the latest technologies but rather a practical working document to aid those who need to know a little of the bacteria that are causing them so much grief, joy or concern.

Why after more than a century of the practice of the science of bacteriology would this be needed? In summary, there are diseases in living systems and there are dysfunctions in natural and engineered systems that can be caused by bacteria. The prime tool for the detection and isolation of bacteria in the last century has been the use of agar-based selective and enriched media. Bacteria that are unable to compete, grow, and be observed on these media are lost as a part of the "false negatives" while attention is paid to the ones that grew on the agar media. There are, therefore, many bacteria that may be very aggressive within natural and pathogenic systems but remain unrecognized as important contributors. Over the last 80 years, microbial ecology has dwelled in the realms of the inconsequential even though it may play essential roles in such events as bioremediation through natural attenuation.

In the last 30 years, the spell-binding unravelling of the various nucleic materials in the bacterial cell has led to a revolution in the precision of identification techniques. This has led to a euphoric embracing of these sophisticated molecular methods of chemical analysis that define cultured strains of bacteria with a high level of confidence. But for this, pure cultures have to be used often grown on agar media. In the "real" world, it is commonly not individual strains that dominate natural events but rather consortia of bacteria functioning within a complex community structure. To identify these structures is therefore a challenge. The Biological Activity Reaction Test (BART™) was co-invented, patented and developed by me as a relatively nonintrusive technique to determine the nature of these consortial structures.

This handbook of bacterial identification has been designed to provide a simple and understandable overview of the identification of bacteria without becoming enmeshed in the linearity of sophisticated chemistry. The emphasis is placed on a general understanding of the traditional principles of bacterial classification and the "nature of being" for selected genera.

This is not intended to be a state-of-the-art document but rather a compendium of the common understandings of the origin of bacterial identification. To achieve this, the book is split into four parts. The first part (Chapter 1) deals with some of the basic terminology practised in bacteriology and relevant to identification. Part two is seven chapters (2 to 8) that deal with the major groups (sections) of bacteria and some of the major component genera in each one are described. Recognizing that within nature, most bacteria function within consortia leads to the third part of the book (Chapter 9) that deals with some identification procedures to define the genera that may be present in particular consortia.

It has often been said that a "picture represents a thousand words" and the fourth part (Chapter 10) incorporates a visual approach to the appreciation of bacterial diversity. To do this, the principles commonly employed by geographers are borrowed to make a bacterial atlas. Here, the bacterial sections become "continents," the families become "countries" and the genera "cities" within which the species form the "suburbs." Thus, all of the sections, families and genera can be easily visualized. Particular features, such as aerobicity, can be illustrated by shading the atlas in a manner that would differentiate the various bacterial groups.

In preparing this book, every effort has been made to "pack" the information into a clear and understandable manner that would aid in the understanding of the identification of bacteria. Like nature itself, the identification of bacteria is beset with variables and this effort represents a methodology to simplify the processes involved. As such, therefore, there are inaccuracies that are unavoidable. However, it is hoped that there is a sufficiently robust information base to begin the bacterial identification understanding. Any identification achieved using this text would have to be confirmed, when desired, using the more sophisticated techniques now available and rapidly being improved.

Introduction

Botanists and zoologists have the luxury of being readily able to view, touch, measure and observe the life cycles of their organisms of interest without any insurmountable difficulty. The net effect is that a vast amount of information can be gleaned from a single specimen while the evaluation of a community rapidly generates an understanding of their relationship to kindred species. For the bacteriologist such luxuries are denied. In its place, the bacteriologist is faced with large communities in which the various component species compete and cooperate.

It is only in recent times that it has become possible to build up an understanding of the nature of these communities (consortia). In nature, consortia of bacteria function within various community structures. These may be summarized as: biofilms, biocolloids, and independent cells functioning within an aqueous matrix (see Figure F.1). By far, the most common growth form is the biofilm which forms as a result of attached growth at a solid-liquid or a liquid-gas interface. These growths rapidly complex causing different bacteria to function within the different strata of the biofilm. Biocolloids may be formed from sheared biofilm fragments, or may be formed within the aqueous matrix by some of the bacteria functioning cooperatively within that matrix. The biocolloid literally "floats" within the aqueous medium and may be able to finely control its density to rise and fall within the water. Other bacteria may function independently as "floating" or motile entities directly within the aqueous matrix (planktonic).

For any microbiologist attempting to isolate and identify bacteria, the challenge begins with the large potential number of organisms that may be present in the sample being investigated. In cell numbers alone, the population can easily reach 1 billion or more cells per mL in a soil or even a clouded water, let alone a pathological specimen. These cells may well exist within a number of community structures and may, or may not, be easily separable into species from each other.

Traditionally, emphasis has always been given to those bacteria which present a major health risk to humans, and animals and plants considered to be of economic significance. Thus the growth of medical and veterinary microbiology along with plant pathology has biased the fundamental methods used to identify bacteria. In pathological specimens, the conditions are often "warped" in favour of the pathogenic bacteria since they commonly dominate the microcosm and are therefore easier to isolate and identify.

This *Practical Atlas for Bacterial Identification* has been developed to allow the users who may be unfamiliar with the sciences involved in microbiology, to begin to comprehend the vast diversity and challenges imposed by the bacterial kingdom upon humankind.

The atlas has four parts. In the first part (Chapter 1) there is a brief introduction to some of the techniques that are commonly applied to obtain a "pure" culture and carry out the primary investigative tasks before attempting identification. In the second part (Chapters 2 to 8), the major groups of bacteria are described. In addition, there are, where appropriate, series of tables that can be used to help understand the traditional ways in which bacteria are speciated in some of the major bacterial genera. Reasons to address the need to identify consortia within natural and infected systems are discussed briefly in Chapter 9.

The final part (Chapter 10) is a display of the diversity and commonality of some characteristics within various groups of bacteria using a "geographical" approach similar to that commonly used in the preparation of an "atlas."

At this time, there continues to be a major debate between various schools of thought interested in the improved classification and identification of bacteria. Such is the extent of this debate that the whole future of the manner in which bacteria are both classified and identified could be significantly changed. In essence the schools of thought range from the idealistic approach of defining the bacterial genome to allow all bacterial classification to be performed at the "ultimate" (molecular) level to the pragmatic approach in which a relatively few simple techniques can be employed to allow an acceptable consensus of an identification. This would involve giving both a genus and species name to the culture with some level of confidence.

The level of acceptable confidence for pathogenic bacteria is, by the nature of the problems created (e.g., infectious diseases), primarily concerned with as accurate and detailed level of identification as possible using all of the tools of molecular biology and science. In contrast, for ecological concerns, there is a need to obtain some understanding of the diversity of the microflora with the emphasis placed more upon its management rather than its composition. The challenge here is the reverse of the problems in medical and veterinary investigations in that the problems are not commonly created by a single species of microorganism but rather by a consortium (often complex) of different species all working together in a synergistic manner (e.g., capitalizing on the diversity for the common good). This book has been written particularly to help guide engineers, consultants, agrologists, and environmental managers in the vast diversity of challenges that these bacteria can create in such diverse places as water wells, oil wells, water systems, heat exchangers, bioremediation, landfill operations and hazardous waste site management. The reader should remember that the microorganisms on this planet are extremely diverse and are capable of growing in environments that we would consider to be extreme. It is not appropriate, although commonly practised, to believe a particular site has too extreme an environment to support microbial growth. It would be more prudent to consider the inevitability of microbial activity if there is liquid water present in any form.

This atlas concept was first presented to students at the University of Regina in 1969 and has, subsequently, gone through eleven major drafts. In developing this concept, the primary goal was to simplify understanding bacterial classification and identification. By presenting the concepts in a visual-spacial manner used in Chapter 10, it has been found that the readers, primarily students, developed a much more confident understanding of the processes of bacterial identification. Thus, just as a geographer can readily illustrate the relative positions of Africa and Europe, so the same techniques can now be employed to illustrate the "relative" positions of the *Enterobacteriaceae* and the *Pseudomonadaceae* to each to other. As the sections of *Bergey's Manual* become "continents" so the families become "countries," the genera "cities" and the species form the sprawling mass of suburbia.

The application of topographical features in the atlas concept is also applicable in the understanding of the relationships between bacterial genera through the process of "shading" the relevant parts of the atlas. For example, the Gram negative bacteria (to the left) can be conveniently separated from the Gram positive bacteria (to the right), the motile (central) from the non-motile (above and below), the aerobes (lower) from the facultative anaerobes (central) and anaerobes (upper), and rods (central) from the cocci (upper right). There are 44 pages of "maps" in this manual. By comparing the various features presented, it becomes possible to rapidly conceptualize the characteristics of the different genera.

Since 1987, I have been involved in the development of the patented Biological Activity Reaction Test biodetection system (BART™, Droycon Bioconcepts Inc., Regina, SK, Canada). This test is based on the principles of reversing redox (oxidative descending to reductive) and nutrient (ascending diffusion) gradients within a selective culture medium to which the bacteria or sample has been inoculated. Reaction patterns generated (colour, location, and form) can be used to develop an initial understanding of the types of bacteria likely to be present in the biodetector as either a pure culture or a consortium from a natural sample. Many of the reactions generated (Appendix) do relate the consortial groups of genera and species which can be typified.

This publication would not have been possible without the forbearance of the many students who suffered through the comprehension of bacterial systematics using this atlas technique. Their many comments, experiences, and observations have allowed this work to evolve. In addition, I would like to acknowledge the laboratory demonstrators, Glenna Seeley and Heather Stanley, who, in turn, have used these principles routinely since 1974. I would like to acknowledge Natalie Ostryzniuk who has, over the last five versions, converted my imperfect text into a manuscript which, hopefully, the readership will now comprehend. The graphics were prepared by Vincent Ostryzniuk using CorelDRAW®. Vincent's ability to produce a much more readable set of atlas pages has contributed in a real way to the usefulness of this text.

D. Roy Cullimore, Ph.D.

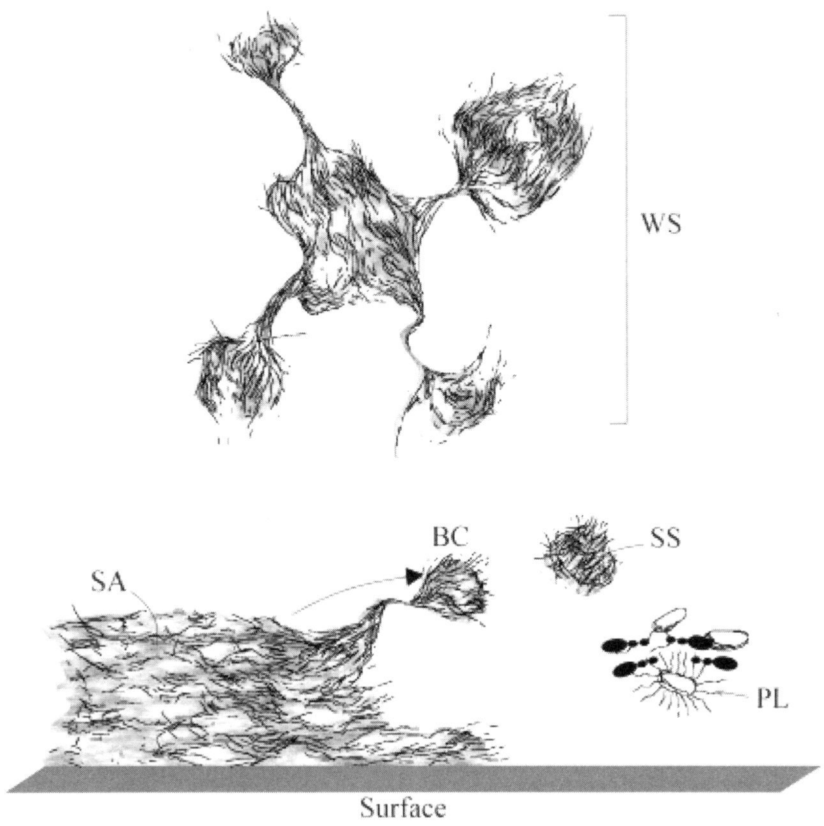

Figure F.1 This figure shows the various forms of growth that will commonly occur in which bacteria will form complex structures. These forms include groupings as sessile attached (SA) as stratified biofilm involving a complex consortium involving many species of microorganisms. When the biofilm sloughs, it can release biocolloidal (BC) particles as suspended "solids" (SS) into the water. This can take the form of "well snow" in water wells. Free-living bacteria are planktonic (PL). Commonly, it can be expected that a natural sample will contain a multiplicity of species commonly in the form of consortia.

About the Author

Dr. Roy Cullimore is an applied microbial ecologist trained at the University of Nottingham, U.K. Since 1975, he has been director of the Regina Water Research Institute at the University of Regina. He, along with co-inventor George Alford, has five patents including the biological activity reaction test (BART™) and the blended chemical heat treatment (BCHT™). Dr. Cullimore has published over 100 refereed papers, 270 technical reports, and has received over $5.5 million in research funding. At present, he is president of Droycon Bioconcepts Inc. of Regina, a biotechnology company involved in research, development, and manufacture. He authored *Practical Manual of Groundwater Microbiology* which was published by Lewis Publishers in 1993. Currently, he is editor of a series of books on sustainable wells. He is now involved in research on the rusticle growths on *RMS Titanic* and dove to the ship in 1996 and 1998 as a part of the Discovery Channel expeditions. Also, Dr. Cullimore is currently involved in the AWWARF water well rehabilitation project being undertaken by Leggette, Brashears & Graham, Connecticut.

Contents

1

Initial Stages to the Identification of a Bacterial Culture

Before starting on the recognition of the various bacterial genera, it is important to gain some understanding of the concepts used. This is a summary of the steps that would commonly need to be taken to ensure that a culture was "pure" (not a contaminated mixture of bacterial strains). This chapter also addresses the application of a suitable set of basic tests applicable to all heterotrophs. This term "heterotrophs" applies to those bacteria able to break down organic materials as a source of carbon and energy.

In practise, one of the first "holy grails" for any bacteriologist wishing to identify a particular bacterial culture is to achieve a pure culture. A pure culture may be loosely defined as a culture of bacteria in which all of the component cells have grown from a single cell, or all of the cells bear a completely common set of characteristics. In other words, the growth (culture) is from a common identical stock of bacteria. Ideally, the culture should consist of bacterial cells, all clones of each other, and bear a complete genetic homogeneity (i.e., all identical). Reality and the rate of mutations (mutagenicity) in bacterial cells make it very difficult to achieve this "holy grail" by the routine methodologies. Therefore, there has to be a practical understanding of what is achievable in the selection of a growth, that it is a "pure" culture suitable for the initiation of the identification protocol.

Traditional protocol requires that the growth be definable by its characteristics. These characteristics are normally more discernable in a colony growing on, or within, an agar-based culture medium rather than in a liquid medium (broth). Such agar media can often contain various chemical and physical agents which aid in the (primarily visual) selection of an appropriate colony for use in the identification protocol. Confirmation that a culture is pure involves evaluation at both the cultural and microscopic levels (Figure 1.1).

At the cultural level, there is a clear advantage in examining the purity using colonial growth over and within the surface of the agar. This is because it is easy to see, recognizable, and relatively simple to sample. To consider a culture to be pure, all of the colonies should possess similar structures, textures, color, and outline. Where this is the case, there may be some differences in the colony types due to the state of maturation of the individual colonies (i.e., differences in size) but all of the colonies should display a common pattern.

Contaminant growth (by other bacterial strains) may be seen by differences (anomalies) between some of the colonies which do not fit into the common pattern displayed by the majority of the colonies. Additionally, where a contaminant strain has grown into a colonial form physically close to a typical colony, there may be one quadrant of the colony that would be different in form. This is a result of the mixed growth of the pure culture with the contaminant infesting that quadrant of the colony.

Broth cultures are more difficult to verify as being pure without the culturing of an agar streak plate to observe the uniformity of the colonial forms. If this is not possible, another technique commonly used is to undertake an extinction dilution in a tenfold stepwise series, in other words a single order of magnitude reduction with each dilution. Eventually the number of cells per mL in the dilution will fall to <1 cell per mL. Where the medium being used to dilute (diluent) is a suitable culture medium, the subsequent incubation of these diluents will cause growth to occur if there are still viable cells in the diluent. Theoretically, there is the highest probability of a pure culture being achieved in the greatest dilution displaying growth after incubation; for example, if a growth culture contained 10^7 cfu per mL and the highest dilution displaying growth was at a 10^{-9} dilution. At this dilution, there would only be a 1 in 100 probability that a cell would be present in any given milliliter of the diluent. If the diluent volume was 10 mL, there would be a 1 in 10 probability of a single cell being present. Therefore, this culture would have the highest probability of being composed of cells from the dominant strain and it would be able to grow in the medium. Microscopic examination of this culture can be used to confirm the uniformity, and hence purity, of the diluent culture.

Microscopic examination can be achieved using a number of techniques. These are described using negative and differential staining techniques. Negative staining, as the name implies, uses a stain that darkens the background and leaves the cell unstained. Indian ink or nigrosin are commonly used for this purpose on either a wet or dried smear of bacterial cells. This enables the observer to determine: (1) the outline shape of the bacterial cell; (2) the presence of a glistening (reflectile) capsule that is intimately associated with the cell itself; (3) the presence of a diffusive extracellular polymeric substance (EPS or "slime") that has generated an ill-defined halo around the cell; and (4) the arrangement of the cells. The advantages of the negative stain are its simplicity (particularly when using a wet mount) and the universality with which cells can be observed by this technique. This latter feature is a distinct advantage since the cells do not have to react in any manner through the direct uptake of stain.

Differential staining can also be used to determine the purity of a culture. The Gram stain forms one of the primary staining techniques which divides the bacteria into two large groups, i.e., pink-red staining is Gram negative; blue-purple staining is Gram positive. Contaminant cells when of the other group can sometimes be distinguished in a Gram stained bacterial smear. Clear edged unstained zones within unevenly stained cells could be formed by the presence of endospores. This should be confirmed by the use of the spore stain in which the spore will be distinctly stained. If the cells are long and slender, often intertwined, the application of the acid-fast stain can be used to determine whether acid-fast cells are present.

Once there has been a reasonable "comfort" that the culture is indeed "pure," then the next stage is to begin to develop a basic data bank which can be used to begin to identify the culture to genus and species. The next three tests routinely used by many bacteriologists are the catalase and oxidase test and the ability of the culture to grow in the presence or absence of oxygen (aerobicity, Figure 1.2).

These three tests can be used as the first step to identifying the culture within a major bacterial group. The usual form of the catalase test is to dip an inoculating needle coated with the culture into a droplet of hydrogen peroxide solution. If there is a foam formation generated around the needle, the culture possesses a catalase enzyme system which has degraded the hydrogen peroxide to oxygen and water. It is the released oxygen that is effervescing to form the foam.

While there are considerable differences in the catalase content of various bacteria, the level of foam formation generated during a catalase test is not usually considered particularly significant. Catalase-positive bacteria are normally aerobic while catalase-negative bacteria are generally either anaerobic or possess alternative mechanisms for coping with the peroxide. It should be remembered that peroxides are commonly produced during aerobic metabolism, and are toxic to bacteria. The catalase enzyme system prevents the build up of peroxide concentrations to toxic levels.

Oxidase is another enzyme system often found in only some of the strictly aerobic bacteria. If a culture has been found to be Gram negative, catalase positive, and strictly aerobic, there is a reasonably high probability that the culture will also be oxidase positive.

Parallel to the other two tests is an evaluation of the aerobicity of the culture. This reflects the ability of the bacterial culture to grow under both aerobic and anaerobic conditions. There are four major groups. Strictly aerobic bacteria will normally only grow in the presence of oxygen and so growth in an agar stab culture is limited to the surface of the agar where oxygen has diffused down into the medium. Some bacteria cannot survive in saturated oxygen conditions but still require oxygen at subsaturation levels. These bacteria are termed microaerobic and commonly function optimally at O_2 concentrations of between 0.02 and 1.5 mg/L (saturation is at 12 to 18 mg/L). In an agar stab culture, growth will therefore occur laterally just under the surface of the agar where oxygen concentrations are reduced.

Facultatively anaerobic bacteria will grow under both aerobic and anaerobic conditions and so the agar stab culture would normally display growth down the full length of the stab. Strictly anaerobic bacteria cannot tolerate the presence of oxygen (which stimulates the production of the toxic peroxides for which they have no defences) and so growth is restricted to the deeper (oxygen-free) depths of the agar stab.

The next stage in the identification is to determine the temperature range over which the culture under investigation would grow. Within the most commonly used temperature range of 0° to 50°C, there are five major groupings of bacteria.

Three temperatures are commonly used to categorize the various bacteria into groupings. These temperatures are 8°, 25° and 37°C. Bacteria can be grouped (Table 1.1) by their ability to grow at these temperatures but it should be noted that there may be a prolonged period of adaptation before a bacterial culture will begin to grow at a temperature outside of the optimal range.

Table 1.1

The Classification of Bacterial Groups on the Basis of Growth Temperatures °C

	8°	25°	37°
Strict psychrotroph	+	-	-
Facultative mesotroph	+	+	-
"Low" mesotroph	-	+	-
"High" mesotroph	-	-/+	+
W.B.A. parasite	-	-/+	+

In traditional practises, the most common incubation temperatures used have been 30°C (warm) and 35° to 37°C (blood heat). The former temperature has been widely used for culturing bacteria from environmental and product sources while the latter temperature has been adopted for the culture of bacterial parasites of warm-blooded animals (W.B.A.) including humans. In the identification of different bacteria, the temperatures of 8°, 25° and 37°C do allow a broad range of bacteria to grow well on at least one of the temperatures. In Figure 1.3, the generalized spectrum of growth for the major bacterial forms is shown. The ability of a culture to grow at one or more of these temperatures allows further classification.

The next stage in the identification of a "pure" culture is to determine the nutritional requirements of the bacteria for growth. Four culture media are routinely used for the routine identification of heterotrophic bacteria. These are the following agar media: glucose ammonium, peptone, glucose tryptone extract, and blood. These media range from simple to complex in that order (see Figure 1.4).

The use of various culture media was initiated when Orla-Jensen, in the early part of this century suggested that bacteria be categorized, in part, on the basis of their ability to utilize combinations of inorganic or organic forms of both carbon and nitrogen. Therefore, the selected media range from a very simple medium which uses glucose as the sole source of carbon and the ammonium ion as the sole source of nitrogen to very complex organic media (e.g., blood agar).

It should be noted that bacteria may take 7 days or longer before displaying growth at some of the temperatures listed previously. Patience is, therefore, a virtue when performing this test!

The simplest medium in the list previously mentioned incorporates a simple organic carbon source (i.e., glucose) and an inorganic form of nitrogen (i.e., ammonium). Bacteria which grows using these compounds are able to synthesize all of the essential organic nitrogenous compounds from the ammonium and do not require any organic sources of nitrogen. This group is referred to as the Orla-Jensen group II.

Group III bacteria all require organic forms of both carbon and nitrogen. They are not able to synthesize all of the necessary sources of nitrogen and, therefore, require some amino acids to be present in the culture medium (i.e., those amino acids that the cells cannot synthesize). The peptone agar is suitable for many Orla-Jensen group III bacteria since it provides a full range of amino acids. However, some group III bacteria require vitamins and special growth factors as well as amino acids for growth. The glucose tryptone extract agar provides a range of vitamins for the growth of the more "fastidious" heterotrophs. Blood agar provides a number of specific growth factors necessary for the growth of some "ultra-fastidious" heterotrophs. Growth on these various media at optimal temperatures for the culture under investigation can be used to determine the nutritional fastidiousness of the bacteria that then helps to determine the probable environments within which the bacteria can be found.

Of all the differential staining procedures, it is the Gram stain which provides the most information and acts as a preliminary tool in the identification procedure. Various forms of data can be gathered using the

Gram stain. These are summarized when spelling the name of the scientist who first reported the technique, Christian Gram. The name "Gram" may be rewritten as g**RAM**. Here, the **RAM** refers to the **R**eaction, **A**rrangement and **M**orphology characterization of the bacteria. Some of the characteristics that may be gleaned are summarized in Figure 1.5.

Reaction refers to the form in which the gRAM stain is taken up by the cell. Beyond the separation of the negative (pink-red) from the positive (purple-blue) reactions, some cells display an unevenness in the stain uptake. These irregular stain patterns occur in both major reactions. These stain patterns include bipolar, barred, and granulose while the rest are evenly stained.

Arrangement is the second major feature in the identification of a bacteria using the gRAM stain. Cells tend to be arranged within common patterns. Some typical forms are as single (independent) cells while others are in pairs, short chains, tetrads, sarcinoids, long chains, small clumps, large clumps and trichomes. In some cultures two, three, or more arrangements may be seen within the same "pure" culture.

Morphology refers primarily to the shape of the cell since the contents are normally homogenous unless inclusions or endospores are present. Care has to be taken in the focussing on a specific cell or group of cells in order to truly view the outline of the cell. The plane of focus of the microscope may only be 0.1 to 0.2 microns. Thus the objective lens has to be positioned so that the outside edge of the cell falls into the sharpest possible focus before an attempt is made to determine the shape of the cell. Common cell forms include spherical (coccoid), flattened coccoid, ellipsoid, coccobacillary, short rod, long rod, and irregular rod. The first five forms often prove to be very difficult to determine without careful observation of a number of the stained cells.

After the gRAM stain, the next technique to pursue is the motility test. This technique commonly uses the hanging drop technique and the maximum motility is often viewed on the edge of the droplet where the oxygen concentration is at its highest. Motility may be viewed by the consistent movement of a cell in a particular direction. Such movement may be linear or erratic and some bacteria may appear to "twitch." Randomized movement (otherwise known as "Brownian movement") is not considered to be a form of motility but is rather the interaction between the cell and the physical forces present within the water medium. Unfortunately, the observation of flagella pattern remains a challenging test to routinely and simply apply to cultures. The form of the flagella pattern remains, however, a major criterion in the identification of bacteria at the species/genus level.

Once the preceding tests have been completed, the further testing which would be required to identify genus and species is variable depending upon the data obtained. The following chapters are written in an order that will take the reader from the more primitive to the more sophisticated genera and then present their relationships through the use of the atlas concept.

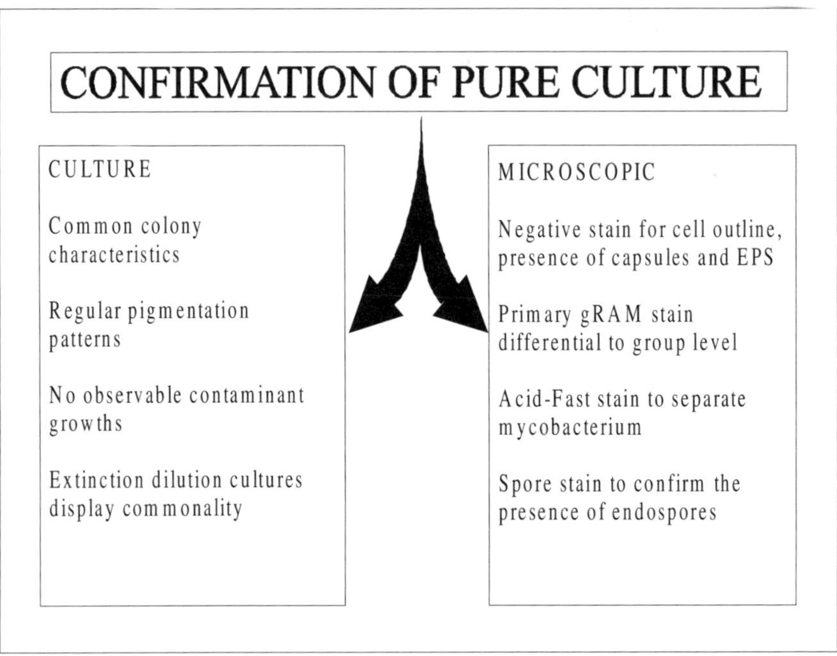

Figure 1.1 The confirmation procedures for beginning to identify a pure bacterial strain involve two approaches at both the cultural (left) and microscopic (right) levels. Cultural practices are often dominated by examining colonies grown on agar media while the microscopic involves a series of staining techniques. It should be remembered that consistent colony characteristics does not mean that the colonies were formed by a single strain of bacterium. Pigmentation of colonies is often delayed until after the colony has grown and can be influenced (positively) by the amount of light. Contaminant colonies would be most recognizable where the colonies are obviously different from the colonies grown from "pure" strains.

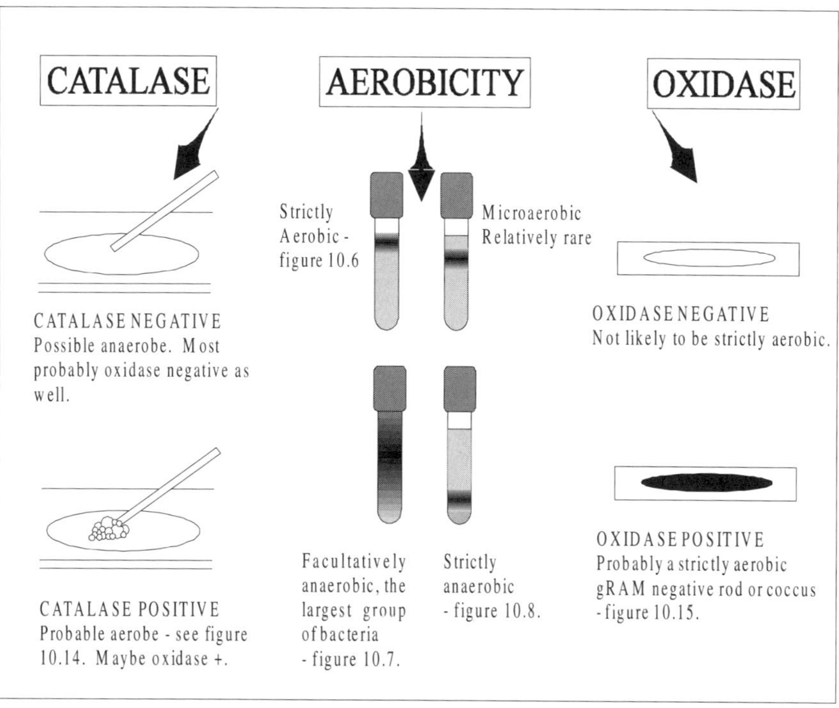

Figure 1.2 There are three basic tests commonly used to start the process of bacterial identification. These include the catalase test (left column), aerobicity test (centre), and oxidase test (right column). Catalase is a test that is almost universally positive for the aerobic bacteria and commonly negative for the anaerobic bacteria. Aerobicity is dependent upon the bacterial strain's ability to grow on the particular agar medium being used for the agar deep stab culture (see Figure 1.4 for more guidance). The oxidase test is most often positive within the aerobic gRAM negative bacteria and is always negative for the strictly anaerobic bacteria.

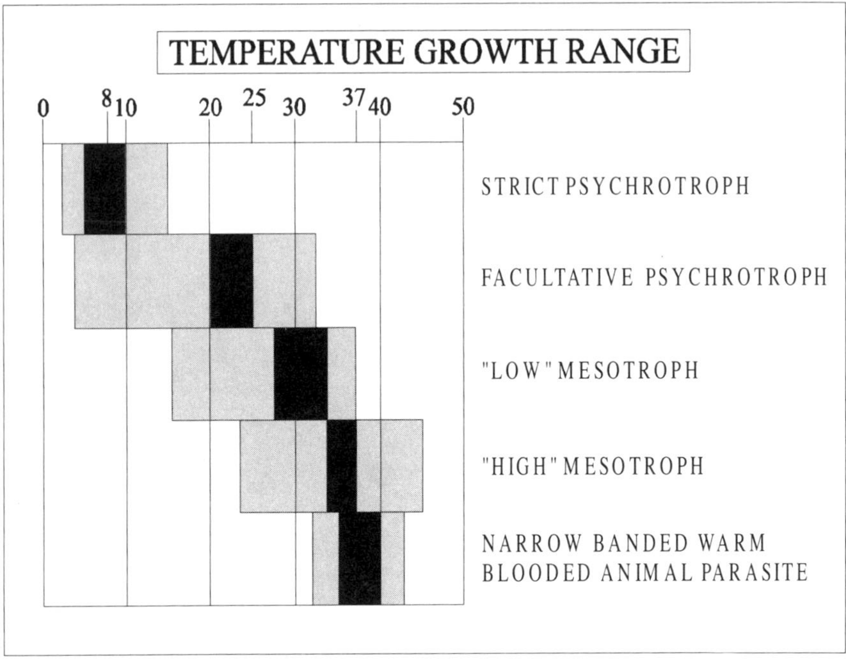

Figure 1.3 Differentiation of the major bacterial groups by temperature growth range. In general, bacteria can be differentiated by the range of temperatures over which they are found to grow. Fastest growth (black bands) is over an optimal range that will vary somewhat for different strains of bacteria. The grey zones show the common temperature ranges in which these various groups of bacteria (described on the right) are able to grow. Thermotrophic bacteria commonly grow under hotter temperatures above 45°C.

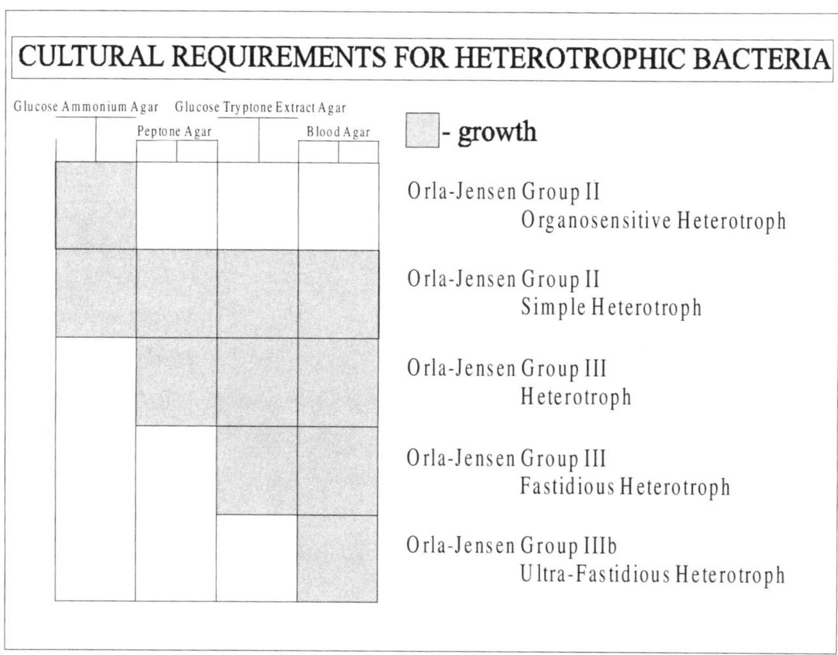

Figure 1.4 Differentiation of bacteria by cultural growth requirements. Four different media are listed as vertical columns. These media range from a very simple glucose ammonium agar (far left) which will support the O-J group II bacteria. Next is the peptone agar (near left) which provides a rich source of amino acids and thus would support many O-J group III bacteria. For the more fastidious O-J III bacteria, the glucose tryptone extract agar (near right) provides vitamins as well as amino acids while the blood agar (far right) provides a broad spectrum of growth factors to encourage the growth of the ultra-fastidious bacteria.

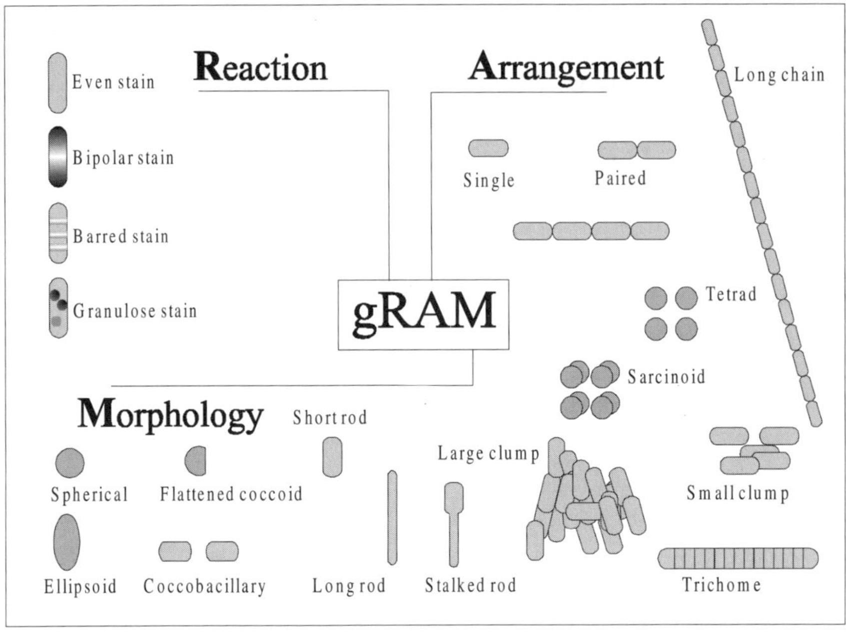

Figure 1.5 gRAM Reaction (upper left), Arrangement (right), Morphology (lower left) characteristics of common types of bacterial cells. Reaction relates to the forms of concentrated staining that can occur within gRAM stained cells. It should be noted that most cells stain evenly. Arrangement relates to the manner in which cells divide and come to form groupings of cells that can be recognizable. Morphology deals with the shape of the individual cells and the most common is rod shaped. Of these characteristics, it is often the arrangement that gives the most "clues" as to the identity of the bacterial strain.

2

gRAM Negative Strictly Aerobic Rods and Cocci (Section 4)

Section 4 bacteria are formed by combining very diverse groups of gRAM negative rods and cocci that share the common feature of being strictly aerobic. This means that the bacteria in this group require oxygen to grow and cannot grow (anaerobically) in the absence of oxygen unless there is a suitable electron acceptor such as nitrate.

Within this section are a range of specialized bacteria that often flourish only within very specific environments. These environments are restricted by the needs for oxygen since the bacteria are all aerobic. Generally, they do not grow in oxygen-free environments (anaerobically) unless there is a suitable alternative electron acceptor such as nitrate.

Differentiation of the Section 4 genera is in two stages (Figure 2.1). The first stage divides the group into eight families, each of which are recognized by some very distinctive features (Figures 2.2, 2.3, 2.4, and 2.5). In the second stage, each of the eight families are divided directly into the component genera. Each family will be introduced separately, the potential pathogens of humans will be described (Figure 2.6), and the genera subsequently will be defined.

Family A, *Pseudomonadaceae*

These strictly aerobic rod-shaped bacteria are catalase positive and normally also oxidase positive. Cells are commonly straight or curved (gRAM negative) rods which when motile (able to "swim") are able to move by polar flagella. These bacteria all respire and are able to use organic carbon compounds as the sole source of energy and carbon.

There are four major genera in the *Pseudomonadaceae* that are separated by distinctive characteristics. These are shown in Table 2.1.

Table 2.1

Identification of the Major Genera in the *Pseudomonadaceae*

	Growth @ pH 3.6	Acid from Glucose	NO_3 red	Oxidase Reaction
Zoogloea	-	-	N_2	+
Xanthomonas	-	+/-	-	+/-
Pseudomonas	-	+/-	var.	+/-
Frateuria	+	+	-	-

N_2, complete denitrification; NO_3, nitrate reduction

Pseudomonas

Of all of the bacterial genera that "crop" up again and again in aerobic environmental problem sites, it is the genus *Pseudomonas* that leads the pack. It is a very diverse genus bearing the common features of the straight to slightly curved rods being rarely nonmotile. The cells may possess a single flagellum or several polar flagella. Lateral flagella are rare. Members of this genera are very adaptable to different carbon substrates and frequently dominate in environments which are aerobic and are presented with a limited range of organic carbon substrates for utilization, e.g., gasoline plume in groundwater. Significant species of interest include *Pseudomonas aeruginosa* (commonly isolated from wound, burn and urinary tract infections), *P. fluorescens* (commonly associated with the spoilage of eggs, cured meats, fish and milk), *P. syringae* (commensal in a range of plant hosts), *P. stutzeri* (nitrate respirer), *P. agarici* (causes drippy gill in mushrooms), and *P. facilis* (able to grow autotrophically using hydrogen). There are three major groups that can be partially speciated using Table 2.2. This table is a summary of a more complex system and allows the primary identification of some of the major species.

Often the members of the genus *Pseudomonas* are there because their characteristics do not fit those used to select the other families and other genera in this family. Hence, it can be considered that this genus is actually a repository for the non-conformist strictly aerobic gRAM negative rods, essentially a "trash" can of convenience!

Xanthomonas

All members of this genus are pathogenic to, or parasitic on, at least one species of plant. They are all straight rods and are motile by a single polar flagellum. The major species is *X. campestris* which possesses a range of strains causing diseases in a wide variety of plants. Within the genus, these plant diseases range from vascular diseases (often causing plugging) through to leaf spot, leaf scald (both of which are severe infections of the leaves), gummosis (slime formation), galls (abnormal growths in the plants) and blights (generalized infections). Some major species can be identified (Table 2.3) using basic test procedures.

Table 2.2
Speciation of Members of *Pseudomonas*

	Pigment	Proteolytic	Starch	Denitrif
aeruginosa	Sol UV Pyo	+	-	+
fluorescens	Sol UV Flor	+	-	+/-
stutzeri	-	-	+	+
alcaligenes	-	-	-	+
flava	Insol y or o	+	+	-

Sol, soluble pigment; UV, UV fluorescent; Pyo, pyocyanin; Flor, fluorescein; Insol, insoluble pigment; y or o, yellow or orange; Starch, hydrolysis of starch; Denitrif, species capable of denitrification.

Table 2.3
Species Differentiation of *Xanthomonas*

	Esculin hydrol	Milk Proteo	Urease	Growth on Nutr agar
campestris	+	+	-	++
fragariae	-	-	-	+
ampelina	-	-	+	+

hydrol, hydrolysis; Proteo, proteolysis; Nutr, nutrient.

Frateuria

This genus differs from the other genera in the family by being able to grow under acidic conditions at a pH of 3.6. This is a trait similar to the acetic acid bacteria family (*Acetobacteriaceae*). The difference is that *Frateuria* is not able to ferment ethanol to acetic acid (the process employed in the manufacture of vinegar). Differentiation of *Frateuria* from the acetic acid bacteria is given in Table 2.4. All members are straight rods that are generally motile by polar flagella but sometimes non-motile phases do occur. This genus is oxidase negative, will grow at a pH of 3.6, can generate H_2S, and will produce acid from ethanol. These bacteria are closely linked to the acetic acid bacteria.

The first three tests relate to oxidation functions while the fourth relates to the ability of the genus to grow on a 30% glucose (gl) medium. The characteristics for *Frateuria* are based on the species *F. aurantia*.

Table 2.4
Differentiation of *Frateuria* from the *Acetobacteriaceae*

	Ethanol to acetic	Acetate to CO_2	Lactate to CO_2	Growth in in 30% gl
Frateuria	-	-	+	+
Gluconobacter	+	-	-	-
Acetobacter	+	+	+	-

Zoogloea

This genus is another very unusual one. The reason for this is that the cells actually use slime threads to connect cells together in a web-like mass (zoogloea). The cells float separately within this web but are held in position by the threads. Cells are commonly straight to slightly curved rods that are often plump and bear the unique characteristic of being able to form zoogloeae. These structures are sometimes involved in the formation of flocs and films on liquid media which can be distinguished by the distinctive "tree-like" or "finger-like" morphologies. Often, *Zoogloea* species are found in organically polluted fresh and wastewaters. Species identification remains to be fully developed but the majority of strains are nitrate respirers, can hydrolyse gelatin and casein, are urease positive, and growth tends to be flocular and concentrated within regions of a broth culture.

Family B, *Azotobacteraceae*

This family has the unique characteristic of being able to fix nitrogen in natural aerobic environments without any parasitism being involved. That means that they can be found in soils, particularly around plants (rhizosphere) and water. These bacteria are strictly aerobic saprophytes which bear the common feature of being able to fixate molecular nitrogen using organic nutrients as the source of energy. Cells are blunt to oval rods often in pairs and chains and can be motile by polar or peritrichous flagella but many are non-motile. There are two major genera, *Azotobacter* and *Azomonas*.

Azotobacter

Azotobacter is a pleomorphic (capable of various cell shapes) dominated by large ovoid cells which may be motile by peritrichous flagella or non-motile. Members of this genus will form a cyst (more protective form of cell) during the life cycle. Ammonium and nitrate can both be used as alternative nitrogen sources and their presence leads to the inhibition of nitrogen fixation. Most strains are both catalase and oxidase positive. Some strains do go through a rod to coccal life cycle during growth. Species are differentiated (Table 2.5) by pigmentation, motility and the ability to reduce nitrate.

Table 2.5

Species Differentiation of *Azotobacter*

	Motile	Pigmentation	NO_3 Red
chroococcum	+	-	+
vinelandii	+	y-gr flor	+
beijerinckii	-	-	+
nigricans	-	br-bl > r-v	+
armenicus	+	br-bl / r-v	-
paspali	+	y-gr flor	-

y, yellow; gr, green; flor, fluorescent; br, brown; bl, black; r, red; v, violet; red, reduction.

Azomonas

Species in *Azomonas* are coccoid, ovoid to rod shaped cells, occur singly, in pairs or small clumps. Cysts are not produced. Motility may be by either polar or peritrichous flagella. Water-soluble and fluorescent pigments are produced by most strains but species differentiation (Table 2.6) is based upon habitat, temperature growth range as well as pigmentation.

Table 2.6

Species Differentiation of *Azomonas*

	Habitat	Growth Range (°C)	Pigment
agilis	w	14 - 37	bl fl*
insignis	w	14 - 32	r-v
macrocytogenes	s	14 - 28	r-v

Pigment, diffusible (soluble); w, water; s, soil; *, only on iron deficient media; r, red; v, violet; bl, black.

Family C, *Rhizobiaceae*

Cells are normally rod shaped and motile by either one polar or subpolar flagellum or 2-6 peritrichous flagella. All species can incite some form of swelling (cortical hypertrophy) in parts of the host plants. These growths may take the form of nodules (on the roots, *Rhizobium*) or galls (on the leaves and stems, *Agrobacterium*) of specific groups of plants. Often, strains can be more easily isolated from the infected galls or nodules than from the surrounding soils. Often the bacteria tend to become less aggressive in the soils and can die out. These bacteria are all aerobic gRAM negative rods that are able to utilize a broad range of carbohydrates and often produce copious extracellular slime formations.

Rhizobium

Rhizobium species are rod shaped cells that easily become pleomorphic under conditions of stress. This genus is characterized by the ability to invade the root hairs of selected leguminous plants (e.g., beans, peas, clover and alfalfa) and incite the formation of nodules on the roots of compliant plants. When cells are present within the nodular structures, they become pleomorphic forming into bacteroids. These now normally become incorporated into that part of the nodule where the active fixation of molecular nitrogen is occurring. The product of this fixation (ammonium) becomes available to the host plant. Species (Table 2.7) are separated by flagella pattern exhibited and the requirement for panto-thenate.

Table 2.7
Species Differentiation of *Rhizobium*

	Flagella Pattern Monotrichous	Peritrichous	Pantothenate Required
leguminosarum	-	+	+
meliloti	-	+	-
loti	+/-	-	-

Agrobacterium

This genus consists mainly of species that generally invade the crown, roots and stems of a variety of dicotyledonous and some gymnospermous plants, via wounds. This causes the invaded host cells to proliferate autonomously into tumor cells. Such induced plant diseases are commonly known as crown gall, hairy root and cane gall. Once the condition is induced, it may be self-proliferating and graftable onto other plants due to the continued presence of these bacterial strains within the plant material. In other words, the plants become infected and the tissues, when grafted, can pass the infection to other plants. The cells are rod shaped and occur singly or in pairs. They are commonly motile by 1 to 6 peritrichous flagella. This genus is not able to fix nitrogen. Species are separated (Table 2.8) by growth in 2% salt, at 35°C, reactions on litmus milk and the presence of urease. Members of this genus are associated with a range of plant diseases and are most commonly linked to the cause of bacterially induced galls. Some species cause such deep-seated infections that the cells automatically enter the seeds to be passed across to the next generation. Some ornamental tobacco plants owe the forms of color in the leaves to infections by these bacteria infecting each and every subsequent generation in exactly the same manner.

Table 2.8

Species Differentiation of *Agrobacterium*

	2% salt*	35°C*	Litmus M	Ure	Oxidase
tumefaciens 1	+	+	Alk	+	+
tumefaciens 2	-	-	Acd	+/-	-
tumefaciens 3	+	-	Alk	-	-
radiobacter	-	-	Acd	-	+
rhizogenes	-	-	Acd	+	+

*, growth; Litmus M, litmus milk; Alk, alkaline reaction; Acd, acid reaction; Ure, urease.

Family D, *Methylococcaceae*

Methane coming up through the crust of the planet from biogenic and thermogenic sources, along with gas generated biogenically in surface organic deposits (e.g., sanitary land fills and sewage lagoons), provides the feedstock for these aerobic bacteria that can utilize these single carbon compounds. They are called the "methanotrophic" bacteria and are formed by a very diverse group of rods, cocci and vibrioids that possess the common ability to utilize methane as the sole source of carbon and energy. This is achieved via the enzyme system, methane mono-oxygenase. There are two primary groups based on whether the strains are obligate methane oxidizers (i.e., only able to use methane) or facultative methane oxidizers (i.e., able to use other carbon compounds as well). All strains are also able to utilize methanol as the sole source of carbon and energy. While formate may be oxidized by some strains, it does not appear that these strains are able to utilize formate as the sole source of carbon and energy unlike methane or methanol. When methane is used as the sole source, it has been observed that complex intracytoplasmic membranes form within the cells. These may be seen as bundles of vesicular disks or paired membranes aligned around the outside of the cell. Strains occur widely in aerobic soils, waters and sediments often adjacent to, or overlaying, anaerobic environments at the redox front. There is one major genus, *Methylomonas*, that has straight rods and produces copious extracellular slime. The cells are most often pink. Both catalase and oxidase positive, these strict aerobes grow optimally where there are saturated or super-saturated oxygen concentrations. Cysts can also be formed.

Family E, *Halobacteriaceae*

It has been commonly thought that bacteria cannot survive in very high concentrations of salt. This has been based, in part, on the traditional practices of pickling foods in brine. The majority of bacteria are inhibited as the salt concentration increases to within the range of 5 to 8%. This family is recognized easily by the common ability of the members to require a minimum of 1.5M (8%) NaCl for survival let alone growth to occur.

Normally, the optimal concentrations for growth range from 3 to 4 M (17 - 23%) of sodium chloride. Cells vary in shape from rods through to cocci and disk shapes but all are gRAM negative. Colonies are often pigmented a shade of red but opaque or translucent colonial forms have also been reported. Most strains are strictly aerobic but some are facultative anaerobes or capable of nitrate respiration. Very commonly, these halobacteria (salt lovers) are found in natural or synthetic environments where there is a high salt (brine) content (e.g., salt ponds, marine salterns and brine preserved fish and meat products). The major genus is *Halobacterium*. Species are differentiated (Table 2.9) by cell shape, ability to use glucose, whether they are alkalinophilic, i.e., able to grow at an optimal pH is 8.5 to 9.5, and also by the ability of the strain to be facultatively anaerobic. Optimal temperature for growth is commonly between 40 and 50°C.

Table 2.9

Species Differentiation of *Halobacterium*

	Fac. Anaer	Cell Disk	NO_3 Reduced	Alkal
salinarum	-	-	-	-
volcanii	-	+	-	-
saccharovorum	-	-	N_2	-
vallismortis	+	-	N_2	-
pharaonis	-	-	-	+

Anaer, facultatively anaerobic; Disk, disk shaped; Alkal, alkalinophilic.

Family F, *Acetobacteraceae*

This family is well known as the acetic acid bacteria since they play a major role in the production of vinegar and the spoilage of beers and wines. The unique feature is the ability of these bacteria to aerobically oxidize ethanol to acetic acid. They are strictly aerobic and can only use oxygen as the terminal electron acceptor. This family is differentiated primarily by the ability of members of this family to oxidize ethanol to acetic acid only in the pH range from 4.5 to 7.5 (*Gluconobacter*), or completely via acetic acid to carbon dioxide (*Acetobacter*). Strains vary from being gRAM negative or gRAM variable with ellipsoidal to rod shapes. Motility can occur by either peritrichous flagella or by 1 to 8 polar flagella but some strains are non-motile. The acetic acid bacteria occur widely in sugary and alcoholized conditions which are slightly acidic (e.g., beer, wine, cider, fruit juices, and honey) and the conditions are aerobic.

Acetobacter

Members of this genus oxidize ethanol to acetic acid but can complete the oxidization of acetate (and lactate) to CO_2 and H_2O. In this genus the cells are ellipsoidal to rod shaped but many involution forms can also occur. Only some strains are motile commonly by either peritrichous or lateral flagella. Most strains do not produce pigments but those which do become pink or brown. Species differentiation (Table 2.10) is based upon the ability to grow on various amino acids (O.J.-III), pigment formation, ability to oxidize 10% ethanol, grow on a variety of carbon sources, and cause ketogenesis from glycerol.

Table 2.10

Species Differentiation of *Acetobacter*

	Sol Pig	Keto	Growth 10% et
acetii	-	+	+
liquifaciens	+	+	+
pasteurianus	-	-	-
hansenii	-	+	-

Sol Pig, water soluble pigment; Keto, ketogenesis; et, ethanol.

Gluconobacter

This is the "twin" genus to *Acetobacter* but is recognized as different by its inability to oxidize acetate. The characteristics used to differentiate this genus from *Acetobacter* include motility (when it occurs, is by 3 to 8 polar flagella), its inability to oxidize acetate or lactate to CO_2 and H_2O, and strong ketogenesis from polyalcohols. Some species are plant pathogens causing such diseases as pink disease in pineapples and a form of rot in apples.

Family G, *Legionellaceae*

A relatively recently discovered genus, this genus gained notoriety through it being identified as the cause of Legionnaire's disease, hence the name. These bacteria are slender rods, usually motile bearing one, two or more straight or curved, lateral or polar flagella. While strictly aerobic, they are relatively fastidious and require L-cysteine-HCl and iron salts for growth. They can utilize amino acids as carbon and energy sources (O.J.-III) but they are not able to oxidize or ferment carbohydrates. They are isolated from surface waters, muds and from hot water systems (e.g., intake and output water pipes to hot water heaters).

Legionella is a very adaptable genus able to survive acidic and alkaline conditions while also having a very broad temperature growth range extending from psychrotrophic through to thermotrophic. It is the only genus in the family and an outline species differentiation is given in Table 2.11. Recognition of the genus is the occurrence of pinpoint colonies on buffered charcoal yeast extract (BCYE) agar after about three days. These colonies enlarge in the next 2 to 4 days at 36 ± 1 °C to appear as 3-4 mm grey glistening, convex, and circular colonies. Most strains produce a brown diffusible pigment on tyrosine containing agar. *L. pneumophila* is the causative agent of legionellosis. This disease may be presented as either a pneumonia with a high fatality rate (Legionnaire's), or as an acute febrile illness (Pontiac disease).

Table 2.11

Species Differentiation of *Legionella*

	Oxidase	Blue flor	Na hipp*	brow.TYE
pneumophila	+	-	+	+
bozemanii	+/-	+	-	+
micdadei	+	-	-	-
dumoffii	-	+	-	+

flor, fluorescence; hipp*, hippurate hydrolysis; brow.TYE, browning of colonies on yeast extract agar with tyrosine.

Family H, *Neisseriaceae*

This family is a small group of coccoid-like aerobic bacteria that bear other characteristics common to Section 4. It is well known because of the presence of *Neisseria gonorrhoeae*, the cause of the sexually transmitted disease, gonorrhea. Almost universally non-motile, some members of this family do possess fimbrae and may exhibit some "twitching" form of motility. Cells are coccoid, commonly single or in pairs or masses. Sometimes the adjacent sides remain flattened along the planes of division. Others are rod shaped or coccoid having only one plane of division. All species are aerobic. All of the genera (Table 2.12) are oxidase positive except *Acinetobacter* and it is presently recognized that oxidase positive species may be parasitic in warm-blooded hosts.

Table 2.12

Differentiation of Genera in Family H, *Neisseriaceae*

	Cell	Oxidase	Catalase	Ferm gl
Neisseria	CFS	+	+/-	Ac/-
Moraxella	R	+	+	-
Branhamella	C	+	+	-
Acinetobacter	C<->R	-	+	Ac/-
Kingella	R	+	-	Ac

Cell, cell shape; CFS, cocci with flattened adjoining sides; R, rod; C, coccus; C<->R, coccus to rod life cycle; Ferm gl, fermentation of glucose; Ac, acidic product.

Neisseria

Species in this genus are commonly cocci occurring either singly or more often in pairs with the adjoining sides abutting the two cells flattened. Division of the cells is in two planes set at right angles and so tetrad forms may be observed. Catalase and oxidase positive, all species produce carbonic anhydrase except the "false neisseriae."

Inhabitants of the mucosal membranes of mammals, some species are primary pathogens for humans. Two major species are *N. gonorrhoeae* (purulent venereal discharges, gonorrhea), and *N. meningitidis* (cerebrospinal meningitidis). It has a tendency to resist decolorization in the gRAM stain procedure. A range of saprophytic species of *Neisseria* and differentiation is given in Table 2.13.

Table 2.13

Species Differentiation of *Neisseria*

	Cell	Arr	Yel Pig	Acid from
gonorrhoeae	C	Pr	-	Gl
meningitidis	C	Pr	-	Gl, Mal
lactamica	C	Pr	+	Gl, Mal, Lac
sicca	C	Pr, Tr	-	Gl, Mal
subflava	C	Pr, Tr	+	Gl, Mal
elongata	R	Pr, Ch	-	-
caviae	C	Pr	-	-

C, coccus; R, rod; Arr, arrangement of cells; Pr, paired; Tr, tetrad; Ch, chains; Yel Pig, yellow pigment; Gl, glucose; Mal, maltose; Lac, lactose.

Moraxella

Species of *Moraxella* are rod shaped but, during growth, the cells becoming very plump in form and approach to, or become, coccoid in shape. They commonly occur in pairs and short chains. The cocci, when observed, are usually smaller and occur singly or in pairs. In the latter event, the adjacent sides are flattened. Cells may be capsulated. Fimbriate cells are sometimes observed having a surface bound "twitching motility." Members of this genus are often parasites on the mucosal membranes of various warm blooded animals including humans. Common characteristics (see also Table 2.14) are oxidase positive, usually catalase positive with glucose not degraded but nitrate may be reduced. While all species are aerobic, some are able to grow under anaerobic conditions.

Table 2.14

Species Differentiation of *Moraxella*

	Cell	Red of NO_3	Heam	Bile st	Phd
lacunata	R	NO_3	-	-	-
bovis	R	-	+	-	-
atlantae	R	-	-	+	-
phenylpyruvica	R	NO_3	-	+	+
catarharrlis	C	NO_3, NO_2	-	-	-
caviae	C	NO_3	+	-	-

R, rod; C, coccoid; NO_3, nitrate reduced; NO_2, nitrite reduced; Heam, haemolysis of human blood in agar; Bile st, stimulation of growth with bile salts; Phd, phenylalanine deaminase.

Acinetobacter

This is an unusual genus in that there is a life cycle in which rod shaped cells become spherical in the stationary growth phase (rod to coccus cycle). Initially, the rods are often plump in form but when they shift to a spherical shape, the cells become smaller than the originating rod shaped cell. There is one major species, *Acinetobacter calcoaceticus,* that is aerobic, catalase positive but, unlike many of the Section 4 bacteria, they are oxidase negative. Most strains are O.J.-II and can grow on simple media using nitrate or ammonium as the sole source of nitrogen. This species occurs widely in waters, sewage, and the soil. Although generally considered nonpathogenic, strains of the acinetobacters have been found to be the causative agents for nosocomial infections in debilitated (e.g., immune compromised) individuals. These infestations range from brain and lung abscesses, through to pneumonia, septicemia, meningitis, to urinary tract infections.

Kingella

Kingella consists of rod shaped organisms that may appear in a paired arrangement, and may be fimbriate. They are only oxidase positive when the tetramethyl reagent is used but, with the dimethyl reagent, this test may be only weakly positive. All species are catalase negative and will produce acid products from the breakdown of glucose. This genus is unusual in that the colonies when growing on an agar surface appear to "corrode" the agar surface as the colonies grow. Sometimes, strains are encountered in samples from the throat, genito-urethral tract, blood and bone but it is generally considered that the pathogenicity is low. Strains appear to be sensitive to penicillin.

Other Genera

There are a range of other genera which have been associated with Section 4 bacteria but have not been placed in any particular family. These genera are peripheral to the basic definition of the families but may remain closely related to one or more of the families. Each genus described below (Table 2.15) possesses unique features as a distinct genus but all bear the common feature that they are strictly aerobic gRAM negative rods or cocci unable to grow anaerobically except through the metabolism of an alternate electron acceptor to oxygen.

Table 2.15

Differentiation of the Other Genera, Section 4

	Oxid	Cat	Mot	Col Pig	Therm	N_2fix
Beijerinckia	+	+/-	-	-	+	+
Derxia	-	-	+	-	-	+
Thermobacterium	-	+	-	-	+	-
Flavobacterium	+	+	-	y	-	-
Alcaligenes	+	+	+	-	-	-
Janthinobacterium	+	+	+	v	-	-
Brucella	+	+	-	-	-	-
Bordetella	+/-	-	+/-	-	-	-
Francisella	-	-	-	-	-	-

Oxid, oxidase; Cat, catalase; Mot, motility; Col Pig, colored pigments; y, yellow; v, violet; Therm, thermophilic; N_2 fix, nitrogen fixation.

Beijerinkia

Members of this genus have slightly curved or straight rods and often occur singly. A unique feature is that molecular nitrogen is fixed under saturated or stressed oxygen regimes commonly in soils. The cells have large highly refractile globules that may occur at one or both poles of the cell. These globules are composed of poly-ß-hydroxybutyrate. Commonly, the cells are capsulated and, unusually, a single capsule may enclose several cells. While cells are often non-motile they can become motile by peritrichous flagella. All are strictly aerobic only using oxygen as the terminal electron acceptor. Growth can occur over a wide pH range from as low as 3.0 to as high as 10.0. In broth media, the whole medium becomes a highly viscous, semi-transparent homogenous mass.

Derxia

This genus is a “close cousin” to *Beijerinckia* but the cells are rod shaped with rounded ends, tending to be pleomorphic. Upon aging, the cells may become filamentous, swollen and distorted with some cells reaching very large sizes (e.g., 30 microns in length). Young cells have an homogenous cytoplasm but as the cells age, refractile bodies appear distributed throughout the cells. Molecular nitrogen is fixed both under conditions of oxygen stress and oxygen saturation. Growth is aerobic and occurs over a narrower pH range from 5.5 to 9.0. This genus is catalase negative. In broth cultures, there is a gelatinous growth which is more luxuriant near the surface and a pellicle is frequently formed at the surface. On agar media, colonies frequently develop a dark mahogany-brown color.

Some strains can grow as facultative hydrogen autotrophs and are most often found in tropical soils.

Flavobacterium

Species of *Flavobacterium* possess one common feature which is that the cells usually develop yellow pigments in colonial growths. The cells are rod shaped with parallel sides and rounded ends, non-motile and do not glide or spread. Cytoplasm inside the cells is homogenous and no granules can be detected. They are all catalase, oxidase and phosphatase positive. A critical characteristic for this genus is that the colonies growing on solid media become pigmented yellow or orange. Acidic products are produced from carbohydrates but no gas in low peptone media. Species are differentiated (Table 2.16) using biochemical characterization. This genus is widely distributed in water and soil and is also often recovered from raw meats, milk and other foods, as well as from the hospital environment and human clinical material.

Table 2.16

Species Differentiation of *Flavobacterium*

	Acid from:	Indole	Casein hydr	Esculin
aquatile	Gl, Suc, L	-	+	-
breve	-	+	+	-
balustinum	Gl, Eth	+	+	+
odoratum	-	-	+	-
mulltivorum	Gl, Suc, Eth, X	-	-	+
spritivorum	Gl, Suc, Eth, L, X	-	-	+

Gl, glucose; Suc, sucrose; Eth, ethanol; L, lactose; X, xylose; hydr, hydroysis.

Thermomicrobium

Members of this genus are short, irregular-shaped rods that are gRAM negative with pleomorphism occurring causing dumbbell or irregular forms to be commonly observed. The major differentiating factor for this genus is the high temperatures at which optimal growth occurs (70 - 75°C) with pH ranges from 7.5 to 8.7. All species are strict aerobic and they are only able to use oxygen as the terminal electron acceptor. Colonies commonly are pigmented with a rose-pink color where good growth occurs on peptone, yeast extract (0.5% of each) agar media. These bacteria are often found in thermal waters such as hot springs.

Alcaligenes

Species are also gRAM negative but occur in a variety of shapes (rods, cocci or cocco-bacillary); the cells usually show an independent arrangement. All species are motile with 1 to 8 (occasionally up to 12) flagella, and they are all oxidase and catalase positive. Carbohydrates are not used but members of this genus are able to use a wide variety of organics and amino acids. Alkali is produced from several of the organic acids and amides and thick slimes may be formed during growth. These species (Table 2.17) occur widely in soil and water but strains have been isolated from clinical material such as blood, urine, feces, purulent ear discharges and wounds. They occasionally cause opportunistic infections in humans. *Alcaligenes* is sometimes difficult to differentiate from some of the other genera of Section 4; the differentiation is given in Table 2.18.

Table 2.17

Species Differentiation of *Acaligenes*

	Respiration using			
	NO_3	NO_2	Gel Hydro	Yel
faecalis	-	+	-	-
denitrificans	+	+	-	-
eutrophus	+	-	-	-
paradoxus	-	-	-	+
latus	-	-	+	-

Gel Hydro, hydrolysis of gelatin; Yel, yellow carotenoid pigment.

Table 2.18

Differentiation of Genera Associated with *Alcaligenes*

	Para/Sap	Str Aer	Alk Lit	Growth Req
Bordetella	P	+	+	Nic
Alcaligenes	S	+	+	-
Brucella	P	+	-	Thi
Heamophilus	P	-	-	X/V
Francisella	P	+	-	Cyst

Para/Sap, parasitic or saprophytic; P, parasitic; S, saprophytic; Str Aer, strictly aerobic; Alk Lit, alkaline reaction on litmus milk; Thi, thiamine; Nic, nicotinamide; Cyst, cysteine.

Serpens

This is a genus that includes the rod shaped cells that extend in length during the stationary growth phase. They may also sometimes develop blebs or spherical protruberances within the cell. The unique feature for *Serpens* is that the cells are extremely flexible and capable of a serpentine motion creeping over agar gel. Cells routinely possess bipolar tufts each of which contains 4 to 10 flagella. There may also be several lateral flagella. All strains are both catalase and oxidase positive. Growth is usually more profuse under conditions of oxygen stress. These are found in eutrophic waters and sediments.

Brucella

This genus includes non-motile cocci, cocco-bacilli or rods that are generally arranged singly. Nitrate can be used as an alternative electron acceptor to oxygen and many strains require supplementary CO_2 for growth and are also O.J.-III and require complex growth media. Blood or serum supplementation is often used to improve growth. All strains are catalase positive while most are also oxidase positive. They are intra-cellular parasites transmissible through a wide range of animals including humans. Of particular concern from the health risk standpoint are *B. melitensis* and *B. abortus.* These cause initially a generalized infection followed by localization in the reticulo-endothelial or reproductive systems respectively. *B. abortus* is a major cause of abortion in cattle.

Bordetella

Species of *Bordetella* are recognized as minute cocco-bacilli that are frequently bipolar, gRAM negative and non-motile. One species is motile by peritrichous flagella. All strains are strictly aerobic with a respiratory metabolism and are fastidious O.J.-III requiring nicotinamide, organic sulfur (e.g., cysteine) and organic nitrogen (amino acids). They are mammalian parasites and pathogens and localized infection sites are frequently among the epithelial cilia of the respiratory tract.

Francisella

This genus includes species that are non-motile and rod shaped in which the cells are very small and become pleomorphic in the stationary growth phase adopting a variety of shapes. All are fastidious O.J.-III and most strains require cysteine for growth. Colonial growth on glucose-cysteine-blood agar gives smooth grey colonies which are formed in 2 to 4 days with a characteristic green discoloration of the surrounding medium. *F. tularensis* is the causative agent for tularemia in humans and animals.

Janthinobacterium

Janthinobacterium species are gRAM negative rods with rounded ends commonly motile bearing both a single polar flagellum and one to four subpolar or lateral flagella. All strains are strictly aerobic and are easily recognized by their ability to produce a violet pigment (violacein). This will cause colonies growing on agar media to be violet in color, while in liquid media, a violet ring appears at the medium/gas interface (pellicle or slime ring). The gRAM stain sometimes reveals a bipolar or barred form of staining with lipid inclusions commonly present. It can grow on ammonium and citrate as the sole sources of carbon and nitrogen and is commonly found in soils and waters but occasionally is associated with food spoilage. This genus is closely related to *Chromobacterium* as the only two genera producing this distinctive violet pigmentation. Differentiation is shown in Table 2.19.

Table 2.19

Differentiation of *Chromobacterium* and *Janthinobacterium*

	Acid products from				HCN	Hydrolysis of		
	Tre	Ara	Xyl	Sor	Prod	Esc	Gel	Cas
J. lividum	-	+	+	+	-	+	-	+
C. violaceum	+	-	-	+/-	+	-	+	-
C. fluviatile	+	-	-	-	-	-	+	-

Tre, trehalose; Ara, arabinose; Xyl, xylose; Sor, sorbitol; Prod, production; Esc, esculin; Gel, gelatin; Cas, casein.

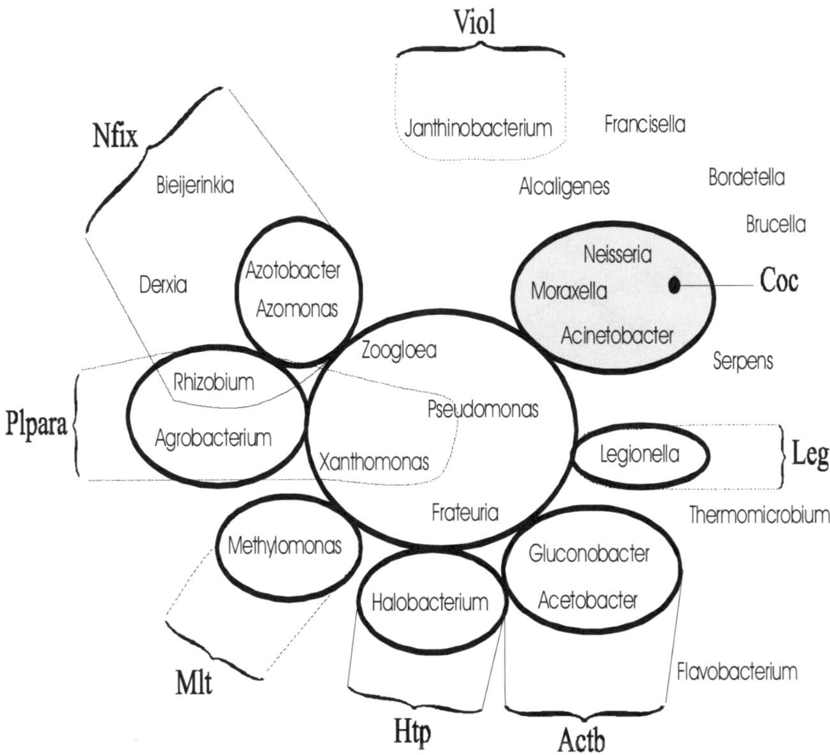

Figure 2.1 Differentiation of the major families in Section 4 based upon major characteristics that include: Nfix, nitrogen fixation; Plpara, plant parasites; Mlt, methylotrophic; Htp, halotrophic; Actb, acetic acid bacteria; Leg, *Legionella*; Coc, coccal forms; and Viol, violacein producers.

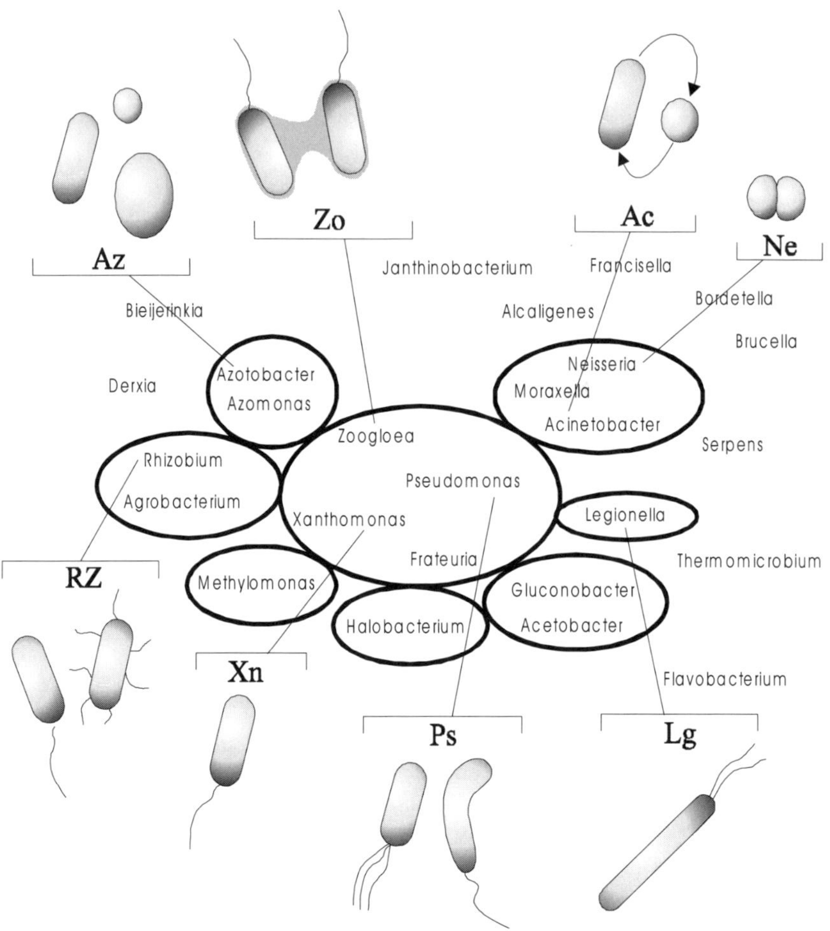

Figure 2.2 Illustration of the various forms of bacterial cells observed in the Section 4. These include: Az, azotobacteria; Alc, *Alcaligenes;* Rz, *Rhizobium*; Zo, *Zoogloea*; Ps, *Pseudomonas*; Xn, *Xanthomonas*; Ne, *Neisseria*; Ac, *Acinetobacter*; and Lg, *Legionella.*

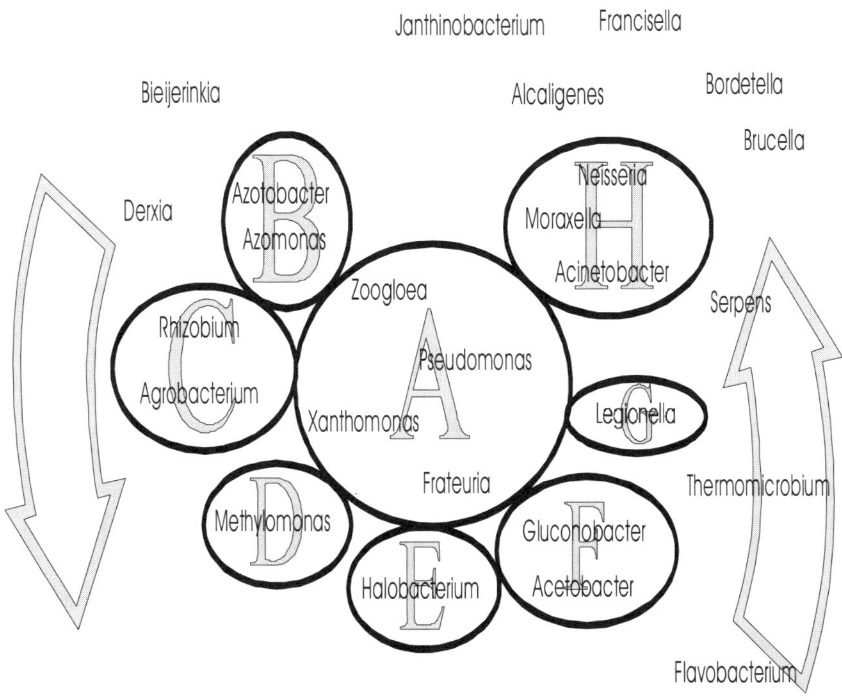

Figure 2.3 Differentiation of the major families in Section 4. These are separated into: A, *Pseudomonadaceae*; B, *Azotobacteraceae*; C, *Rhizobiaceae*; D, *Methylococcaceae*; E, *Halobactiaceae*; F, *Acetobacteraceae*; G, *Legionellaceae*; and H, *Neisseriaceae*.

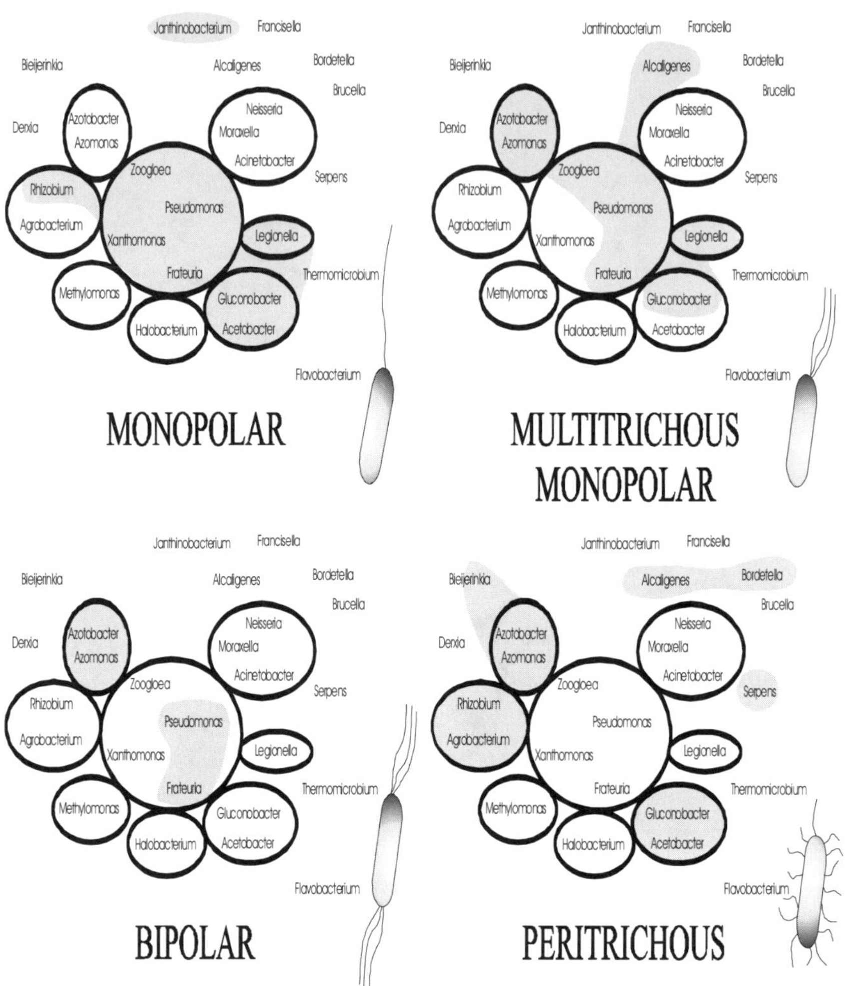

Figure 2.4 Flagella patterns commonly occurring in the various genera in Section 4. There are four boxes which give, by shade, the genera possessing the specified type of flagella pattern. This pattern is both named: single flagellum monopolar (upper left), multitrichous monopolar (upper right), bipolar (lower left) and peritrichous (lower right).

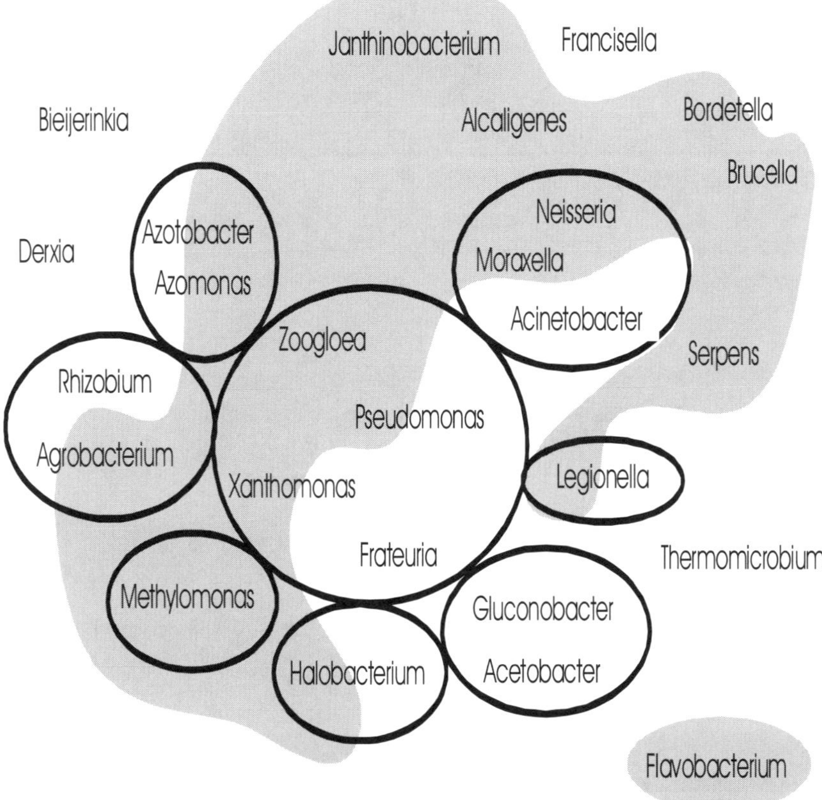

Oxidase Positive

Figure 2.5 Diagram illustrating the genera in Section 4 that possess the oxidase enzyme system. Genera in which all species are oxidase positive are shaded, those genera in which only some species are oxidase positive are presented by a partial of shading of the name. Oxidase negative genera are not shaded.

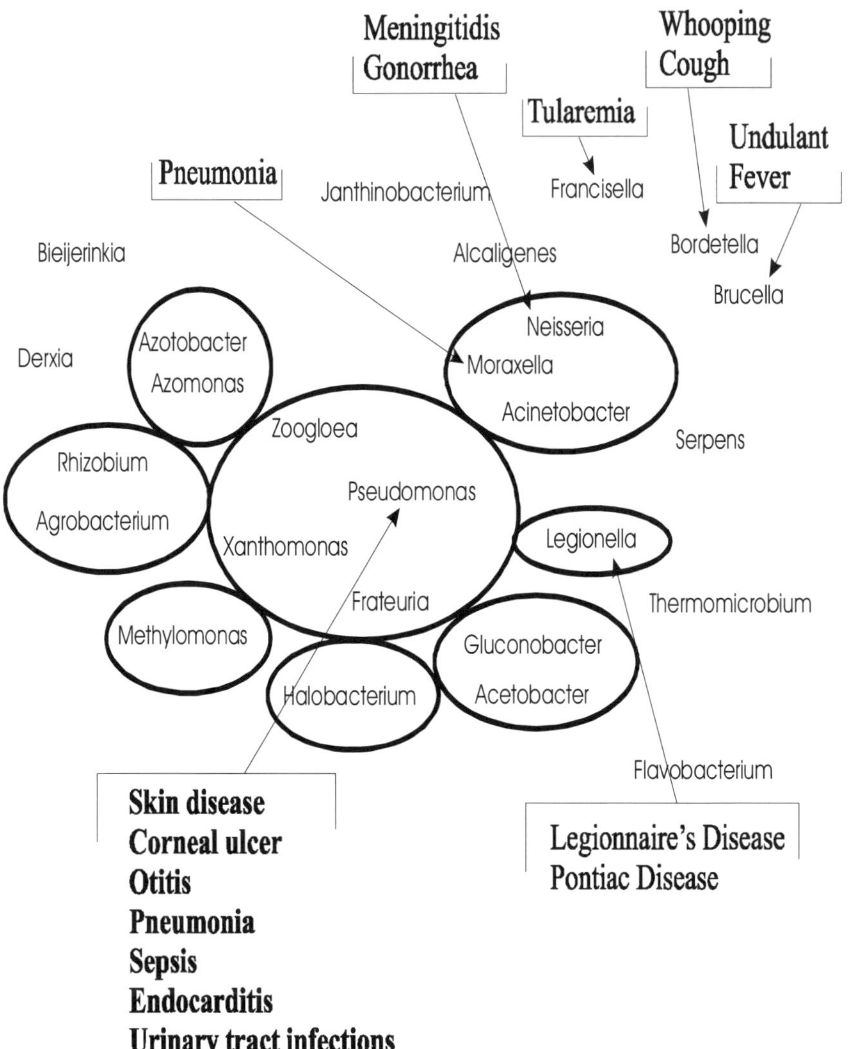

Figure 2.6 Diseases in humans caused by pathogenic species of Section 4 bacteria. A single bracket is used for each genus in which some species that are pathogenic occur. The bracket uses an arrow to indicate the genus and the various diseases are listed within the box.

3

Facultatively Anaerobic gRAM Negative Rods (Section 5)

While Section 4 bacteria are all strictly aerobic, Section 5 are all facultatively anaerobic. There are three families (Table 3.1, Figure 3.1) separated by whether they are motile, non-motile, oxidase positive or oxidase negative. They are all gRAM negative catalase positive, facultatively anaerobic rods. This group includes a number of major human, animal and plant pathogens and also forms a major focus for the indicator organisms associated with hygiene risk (i.e., the coliform bacteria, Figure 3.2).

Table 3.1
Differentiation of the Three Families of the Section 5 Bacteria

	Oxidase	Motility
Enterobacteriaceae	-	+
Vibrionaceae	+	+
Pasteurellaceae	+	-
Motility, see also Figure 3.3 for flagella patterns.		

Family A, Section 5, *Enterobacteriaceae*

This family is most known for the fact that the "enteric" bacteria form a part of this family because of the importance of the coliform bacteria and the significance of typhoid, food infections (salmonellosis), and bacterial dysentery (shigellosis). See also Figures 3.4 and 3.5. There are many other genera of significance as plant pathogens, causes of food spoilage, and of major importance in the environment. While they can be split into five "tribes" (Table 3.2), the *Enterobacteriaceae* are all gRAM negative rods that are catalase positive and, where motile, possess peritrichous flagella. Most genera will reduce nitrate to nitrite (denitrification). Some of this family are simple chemoorganotrophs (O.J-II) using D-glucose as the sole source of carbon and energy, while others are very fastidious requiring various vitamins and amino acids. All can grow in the presence or absence of oxygen. Each of these five tribes will be addressed separately with emphasis placed on the differentiation of the genera within each tribe.

Table 3.2

Differentiation of the Five Tribes of the *Enterobacteriaceae*

Tribe #	Habitat
I	Primarily parasitic in animals
II	Primarily saprophytic, some can be pathogenic
III	Saprophytic
IV	Plant pathogens
V	Pathogenic using vectors for transmission

Tribe I, consists of five genera, all of which are primarily parasitic in animals with some causing serious diseases particularly in humans. The genera (Table 3.3) include: *Escherichia, Edwardsiella, Shigella, Salmonella* and *Citrobacter*. *Escherichia* includes many relatively benign strains that parasitize the intestine (gastro-enteric tract) of many animals and are particularly dominant in the human species. Some are opportunistic pathogens and can cause forms of diarrhea. *Edwardsiella* is a common parasitic genus found in reptiles including snakes, toads and frogs. There are many similarities between *Escherichia* and *Edwardsiella* that make differentiation challenging. *Shigella* is an intestinal pathogen causing bacillary dysentery. *Salmonella* causes three major infections related to enteric fevers (typhoid, paratyphoid), food infections that can cause gastro-enteritis, and septicaemia. *Citrobacter* is essentially a "border line" genus with the tribe II and includes a range of opportunistic pathogens and also some are commonly found in the environment.

Table 3.3

Differentiation of Genera in Tribe I, *Enterobacteriaceae*

	Cit	H_2S	Ldc	Mot	KCN	Glgas
Escherichia	-	-	+/-	+	+/-	+
Edwardsiella	-	+	+	+	-	+
Citrobacter	+	+/-	-	+	+	+
Salmonella	+	+/-	+	+	+	+
Shigella	-	-	-	-	-	-

Cit, citrate utilized; H_2S, hydrogen sulfide produced; Ldc, Lysine decarboxylase; Mot, motility; KCN, utilizes potassium cyanide; Glgas, gas produced during fermentation of glucose.

Tribe II, includes genera that are primarily saprophytes found commonly in the environment although some can be pathogenic (Table 3.4).

Table 3.4
Differentiation of Genera in Tribe II, *Enterobacteriaceae*

	Prod	Caps	Cit	Ind	Mot
Enterobacter	-	-	+	-	+
Serratia	+	-	+	+	+
Klebsiella	-	+	+	+/-	-
Hafnia	-	-	-	-	+

Prod, prodigiosin (red pigment); Caps, capsulation; Cit, citrate utilized; Ind, indole produced; Mot, motility.

This includes four genera: *Enterobacter* is a widespread genus that can be found in soils, muds, waters and organic spoilage but is also commonly found in animals and humans. *Serratia* also occurs widely in the environment but some are opportunistic human pathogens. *Hafnia* is commonly found in human, animal and bird feces but is also common in organically polluted waters and soils. *Klebsiella* is widely distributed in soils, waters, woody plants and the intestine but is most recognized for its ability to cause bacterial pneumonia.

Tribe III, includes three genera (Table 3.5) that are tied together by the common possession of the phenylalanine deaminase enzyme system. *Proteus* occurs in a variety of organically polluted waters, soils and manures but also can be pathogenic causing urinary tract infections or septic lesions. *Providencia* is commonly found as an animal parasite often isolated from stools, urinary tract infections, bacteremia and burns. It does not appear, however, to be pathogenic. *Morganella* is found routinely in feces and has been found to be an opportunistic pathogen causing both respiratory and urinary tract infections.

Tribe IV, contains the single genus, *Erwinia*, that is recognized as a common part of the microflora on plants (epiphytes) and also as a plant pathogen. Some strains do, however, grow saprophytically.

Tribe V, also contains only one genus, *Yersinia,* which has been found in a wide variety of habitats. Some are pathogenic and vectors can be involved in the spread of the disease.

Coliform bacteria are recognized to be of major importance in the health risk assessment of water and waste waters. There are two major groupings referred to as the total and fecal coliforms. While both indicator systems are in common usage, it is the total coliforms that give a broader "sweep" of the potential coliform bacteria while the fecal coliforms are "focussed" tightly on coliform bacteria that could have originated from fecal material. The genera commonly recovered in coliform tests are listed in Table 3.6. The description of the various genera in the *Enterobacteriaceae* is reported below by tribe in order from I through V. Differentiation of the tribes to genus is given where relevant.

Table 3.5

Differentiation of Genera in Tribe III, *Enterobacteriaceae*

	V.P.	Odc	Gel	Cit	Swm
Proteus	+/-	+/-	+	+/-	+
Providencia	-	-	-	+	-
Morganella	-	+	-	-	-

V.P.,Voges-Proskauer; Odc, Ornithine decarboxylase; Gel, gelatinase; Cit, citrate utilized; Swm, swarms on agar surfaces.

Table 3.6

Enterobacteriaceae Detected in the Coliform Tests

	Fecal Coliform	Total Coliforms
All species	*Escherichia*	*Escherichia*
Most species		*Enterobacter*
		Klebsiella
Some species	*Enterobacter*	*Serratia*
	Klebsiella	*Proteus*
Non-enterics		*Aeromonas*
		Pseudomonas

Most species would mean that the likelihood of presence is in the range of 40 to 80% of incidences whereas for "some" species, the likelihood of occurrence is less than 40%. Non-enterics are in either of these ranges.

Escherichia

This genus is the prime indicator for hygiene risk in waters and foodstuffs. This is because *Escherichia coli* occurs in very high numbers in human feces and their presence in water, beverages and food indicates a health risk to the consumer. The cells in this genus are straight rods occurring singly or in pairs and motility is very common with peritrichous flagella. Occasionally, capsules or micro-capsules may be present and slime (thin) or mucoid (thick) production can occur. Glucose and some other carbohydrates are fermented with the production of pyruvate which is then converted into lactic, formic and acetic acids. Part of the formic acid is split by a hydrogenylase system into CO_2 and H_2 in equal amounts. It is, therefore, common for this genus to generate both an acid and gas reaction when fermenting sugars. Lactose is also fermented by most strains. Strains of *Escherichia* occur commonly in the lower part of the intestines of warm blooded animals including humans. *E. coli* is an opportunistic pathogen and some strains can cause diarrhea (e.g., entero-pathogenic *E.coli I,* EPEC, and entero-toxigenic *E.coli,* ETEC). The strain of *E.coli* which is of considerable concern today is strain 0157.

Edwardsiella

This genus can be considered the "twin" of *Escherichia* except that it dominates in the intestines of many reptiles including frogs, toads and snakes. Cells are small straight rods and are motile by peritrichous flagella. These are all O.J.-III and require both amino acids and vitamins for growth. D-glucose is fermented with the production of acid and sometimes small volumes of gas. This genus is not able to ferment as wide a range of carbohydrates as other members of the family. Colonial growth on peptone agar media is weak and only small diameter (0.5 - 1.0 mm) colonies develop.

Citrobacter

Citrobacter can be viewed in some ways as the environmental "twin" of *Salmonella.* Cells in this genus are straight rods that are usually motile by peritrichous flagella. Citrate can be used as the sole source of carbon (hence the prefix for the name). Glucose is fermented to acid and gas. While often isolated from clinical specimens as opportunistic pathogens, strains are also commonly found in soil, water, waste waters and food.

Salmonella

This major genus forms one of the bacterial groups that present a range of serious health risks to humankind. This is because many species are able to infect the body via the intestine and cause a range of diseases from enteric fevers, gastro-enteritis and septicemia. The most serious form is typhoid caused by *Salmonella typhi.* Cells are all straight rods and usually motile with peritrichous flagella. Gas is usually produced from glucose while H_2S is generated in triple sugar iron agar by many strains. Like *Citrobacter*, all strains can use citrate as the sole source of carbon and also are able to infect many animal species. Strains are identified primarily using serological techniques.

Shigella

Species in this genus are notorious as one of the more severe causes of dysentery. It differs from other members of Tribe I in that the cells are non-motile. Typical cells are, however, straight rods that are able to ferment sugars but without gas production. Unlike some other genera in this tribe, *Shigella* species cannot utilize citrate or malonate, grow in KCN or produce H_2S. All species are intestinal pathogens causing bacillary dysentery in primates and humans.

Enterobacter

A very widely distributed and adaptable member of tribe II and is found occurring in soils, waters, plants and any spoilage of organic materials as well as in animals and humans. Cells are straight rods and motile by sparsely peritrichous flagella (normally 4 to 6 flagella are present). Glucose is fermented to acid and gas (moderate volume). Glucose is not fermented to gas at 44.5°C. Citrate or malonate can be used as the sole source of carbon. They are Voges-Proskauer positive. H_2S is not produced from thiosulfate. Proteolysis of gelatin occurs slowly. Two major species are *E. cloacae* and *E. aerogenes*. Differentiation of these species is given in Table 3.7.

Table 3.7

Species differentiation in *Enterobacter*

	Odc	Ldc
cloacae	+	-
agglomerans	-	-
aerogenes	+	+

Odc, ornithine decarboxylase; Ldc, Lysine decarboxylase.

Serratia

This genus stands out in Tribe II because it commonly possesses a red distinctive pigment in colonies growing on many agars. The cells are straight rods with rounded ends normally motile by peritrichous flagella. Colonies are often opaque but somewhat iridescent and may shift during growth from a white to a pink or red color. This is due to the production of prodigiosin and pyrimine. Prodigiosin is a non-diffusible water insoluble pigment. Of the various agar media, peptone-glycerol agar can be used to most reliably generate this pigment. Matured colonies may be totally red or show either a red center, red edge or, at least, a red sector. Pyrimine is a water soluble diffusible pink pigment which will diffuse out into the agar to produce a pink halo around the colony. Sometimes, the whole agar medium will turn pink. Catalase is very strongly positive. Lipase and protease exoenzymes are normally produced but not amylase. Cultures can produce two kinds of odors, a strong fishy/urinary odor or a musty potato-like odor. These organisms occur widely in the environment but some are opportunistic human pathogens. There are three major species that are differentiated in Table 3.8.

Table 3.8
Species Differentiation in *Serratia*

	Growth @ 4° C	Ldc
marcescens	-	+
liquifaciens	+	+
odifera	+	-

Ldc, lysine decarboxylase.

Klebsiella

Another very ubiquitous member of Tribe II, this genus is well known as the cause of bacterial pneumonia but little is recognized of the other species of this genus that occur in a very wide variety of habitats. It is the protective nature of the capsule that surrounds the cells in this genus that provides an "edge" in competition with other bacterial genera. Cells are non-motile straight rods occurring singly, in pairs or short chains. They are easily recognized by the formation of capsules which on meat extract agar media can cause the colonies to become domed, glistening and "sticky." This genus has simple nutritional requirements for growth (O.J.-II) and normally both glucose and citrate can be used as the sole source of carbon. Glucose is fermented with the production of acid and gas in small volumes. The differentiation of some of the major species is listed in Table 3.9. There is a wide variation in the genus and some strains are able to fix molecular nitrogen. Widely distributed in nature, strains are found in soils, waters, grain, woody plants, clinical specimens and intestinal contents. *K. pneumoniae* has been isolated from several pathological conditions in humans particularly infections of the urinary and respiratory tracts. Capsule types 1, 2, and 3 may be the causative agent for pneumonia.

Table 3.9
Species Differentiation in *Klebsiella*

	Growth @10°C	M.R.	V.P.	Ind
pneumoniae	-	-	+	-
ozanae	-	+	-	-
oxytoca	+	-	+	+
terrigena	+	+	+	-

M.R., methyl red test; V.P., Voges-Proskauer test; Ind, Indole.

Hafnia

This genus occurs widely in the feces of humans and other animals including birds. The cells are straight rods and motile by peritrichous flagella but non-motile strains also occur. Most strains can use citrate, acetate or malonate as the sole source of carbon. Nitrate is reduced to nitrite. Glucose is fermented to acid and gas. Members of this genus are also commonly recovered from sewage, waters, wastewaters, soils and dairy products.

Proteus

Species belonging to *Proteus* are most often recognized by their ability to display periodic cycles of "swarming" over agar plates. Swarming is a slime-like colonial growth occurring rapidly over the agar surface as a regular or irregular growth. This phenomenon involves the cells spreading out over moist surfaces (such as the surface of an agar or gelatin medium) to produce a series of broken or entire concentric rings or a continuous film of growth. It is troublesome for bacteriologists trying to separate different colony types using the agar spread plate. The cells are straight rods capable of motility by peritrichous flagella. Another unique feature is the ability of strains to deaminate phenylalanine and tryptophane. Urea is hydrolysed and acids are produced from mono- and di-saccharides but cannot ferment straight chain tetra-, penta-, or hexahydroxy alcohols. Hydrogen sulfide is produced. Some strains are pathogenic causing urinary tract infections but also secondary infections can be caused leading to septic lesions at various sites in the body (e.g., ear infections, otitis). Strains also occur widely in polluted (organically) waters, soils and manures and the most common species is *Proteus vulgaris*.

Providencia

This genus is related to *Proteus* and while it is motile by peritrichous flagella, it is not able to swarm. Cells are straight rods that are able to produce acid from one or more of the polyhydric alcohols. Acids are also produced from mannose and citrate. Generally, it can be isolated from diarrhetic stools, urinary tract infections, bacteraemia and burns but any role in disease production has not yet been firmly established.

Morganella

This is another genus that is also closely related to *Proteus* and *Providencia.* The cells are straight rods often motile by peritrichous flagella at lower temperatures. While a surface film of growth may form on agar media, an aggressive swarming does not occur. Few carbohydrates are fermented and citrate is not utilized but tartrate is. Phenylalanine and tryptophane are deaminated oxidatively. Strains are routinely isolated from the feces of many mammals including humans. It can be a secondary opportunistic pathogen and strains have been isolated from respiratory and urinary tract infections and bacteraemia.

Erwinia

This is the only genus in Tribe IV. While some are saprophytic, many form a part of the epiphytic flora of plants or are plant pathogens. Plant diseases are diverse and include fire blight (*E. amylovora*), vascular wilt (*E. tracheophila*), leaf spot (*E. maltivora*), phloem necrosis (*E. rubrifaciens*), and rotting (*E. carotovora*). Species differentiation is given in Table 3.10. Cells are straight rods that are motile by peritrichous flagella. Cells occur singly, in pairs or short chains. Like other Section 5 bacteria, these bacteria are facultatively anaerobic but growth under anaerobic conditions is sometimes very weak. Acid is produced from some carbohydrates including fructose, D-glucose, galactose and sucrose. It can also utilize acetate, fumarate, gluconate, malate and succinate.

Table 3.10

Species Differentiation of *Erwinia*

	O.J.	H_2S	NO_3	Gel	Pect.	36°C
amylovora	IIIf	-	-	+	-	-
tracheophila	IIIf	+	-	-	-	-
herbicola	III	+	+	+	-	+
carotovora	III	+	+	+	+	+
maltovora	IIIf	-	-	-	-	-

O.J., Orla-Jensen nutritional grouping; IIIf, fastidious; H_2S, hydrogen sulfide produced; NO_3, nitrate reduced to nitrite; Gel, gelatinase; Pect, pectinolytic; 36°C, growth at 36°C.

Yersinia

The only genus in Tribe V, strains occur in a wide range of habitats (animate and inanimate) with some strains adapted to specific conditions or hosts including the use of various animal vectors in the infection process. *Y. pestis* is the causative agent for plague, while *Y. enterocolitica* is responsible for forms of diarrhea, terminal ileitis, mesenteric lymphadenitis, arthritis and septicemia in animals including humans. Cells are cocco-bacillary or short rods. They are non-motile at higher growth temperatures (e.g., 37°C) but are motile at lower temperatures with peritrichous flagella (an exception is *Y. pestis* which is non-motile). Capsules are not formed although *Y. pestis* will produce an envelope around the cell. Glucose and other carbohydrates are fermented with acid production but rarely gas (small volume). Phenotypic characteristics are influenced by the incubation temperature at which the strain is grown and 25° to 29°C are the common conditions used to characterize strains.

Family B, Section 5, *Vibrionaceae*

The major defining feature of this family is that the genera in this family are oxidase positive motile facultative anaerobes. Cells are commonly straight to curved rods commonly motile by polar flagella (rather than the peritrichous forms seen in family A). Occasionally, lateral flagella may also be present but these have a different wavelength to the polar flagella. Most strains can utilize D-glucose as the sole source of carbon and energy. In addition, ammonium is commonly used as the sole source of nitrogen (O.J.-II). In general, strains have a simple nutritional requirement but many require 2 to 3% NaCl in the culture media in order to grow. These organisms primarily live in aquatic habitats. Several species are, however, pathogenic for fish, eels, and frogs as well as other invertebrates and vertebrates including humans. There are four genera in the family and are differentiated in Table 3.11.

Table 3.11

Differentiation of the Genera in Family B, *Vibrionaceae*

	Cell shape	Biol	0/129	Oxid
Vibrio	Sr, Cr	-/+	+/-	+
Aeromonas	R-C	-	-	+
Photobacterium	Sr	+	-	-
Plesiomonas	Rre, f	-	+	+

Sr, straight rod; Cr, curved rod; R-C, rod and coccoid forms; Rre, rods with rounded end; f, filamentous; Biol, bioluminescent; 0/129, sensitive to Vibriostatin 0/129; Oxid, oxidase.

Vibrio

A very large and diverse genus, it is found in a wide range of aquatic environments and is very common in estuarine and marine waters particularly at lower temperatures. It is also found upon and within the bodies of marine animals. Some species are pathogenic to marine animals and humans. Most significant is *V. cholerae* which is the pathogen causing cholera. Cells are straight or curved rods which are motile by mono- or multi-trichous polar flagella. In liquid media, these polar flagella are unusual in that they can be enclosed in a sheath continuous with the outer membrane of the cell wall. On solid media, lateral flagella may also form. Oxygen is the universal electron acceptor and nitrate cannot be used. As simple chemoorganotrophs (O.J.-II), they are able to grow on mineral salts media containing D-glucose and NH_4Cl. Most strains require elevated levels of the sodium ion for growth (minimal critical conc., 5 - 700 mM). D-glucose, D-fructose, maltose, and glycerol are fermented to acid but normally without gas. *V. anguillarum* has been associated with a range of diseases in fish. Species differentiation is given in Table 3.12.

Table 3.12

Differentiation of the Species of *Vibrio*

	Biol	40°C	Util	Swrm
cholerae	-	-	S.,G.	-
harveyii	+/-	+	s,C,G	-
anguillarum	-	-	S,c,g	-
fischeri	+	-	C	+
proteolyticus	-	+	G	-

Biol, bioluminescent; 40°C, growth at that temperature; Util, utilization of (upper case, all strains; lower case, some strains only); S, sucrose; C, cellobiose; G, glucose; Swrm, swarming on solid complex media.

Photobacterium

Species of *Photobacterium* are found growing on the surfaces and in the intestines of marine animals. Some are symbionts in the specialized bioluminescent organelles of marine fish (e.g., *P. phosphoreum*). In the deep oceanic environments, the blue glow from the deep scattering layer and the light glowing out of specific parts of marine animals are often due to the active presence of this genus. Cells are plump straight rods which accumulate poly-ß-hydroxybutyrate. Distorted (involution) cell forms are common in older cultures. Some strains are motile by one to three polar flagella but some strains are non-motile. Simple O.J.-II chemoorganotrophs, they are able to grow in simple mineral salts media with D-glucose and NH_4Cl. Some strains do, however, require L-methionine for growth. Sodium ions are commonly required for growth. Strains are found in marine environments and two species are bioluminescent. There are three species differentiated in Table 3.13.

Table 3.13

Species Differentiation of *Photobacterium*

	Ggas	Biol	35°C	Lipo	Util
phosphoreum	+	+	-	-	M
leignathi	-	+	+	+	A
angustum	-	-	+	+/-	X,M,a

Ggas, gas from glucose; Biol, bioluminescent; 35°C, growth at; Lipo, lipolysis; Util, utilization of (upper case, all strains; lower case, some strains only); M, maltose; A, acetate; X, D-xylose.

Aeromonas

This species occurs widely in freshwater but some are pathogenic to fish and frogs. For example, *A. salmonicida* causes furunculosis in salmonid fish, and *A. hydrophila* can be pathogenic to frogs, fish and mammals including humans. Cells may be rod shaped with rounded ends or coccoid occurring singly, in pairs or short chains. Motility is by a single polar flagellum but peritrichous flagella may be found in young cultures growing on agar media. Carbohydrates are broken down to acid or acid and gas. All are oxidase positive and resistant to the vibriostatic agent 0/129. Species are differentiated in Table 3.14. Strains can use a variety of carbohydrates and organic acids as the sole source of carbon.

Table 3.14

Species Differentiation of *Aeromonas*

	Mot	Cell	Esc	V.P.	H_2S
hydrophila	+	Cb,R	+	+	+
caviae	+	R	+	-	-
salmonicida	-	Cb	+	-	-
achromogenes	-	Cb	-	-	-

Mot, motility; Cell, cell shape; Cb, coccobacillary; R, rod; Esc, esculin; V.P., Voges-Proskauer; H_2S, from cysteine.

Plesiomonas

Species of this genus are generally believed to occur normally in the intestines of humans and have been found to be a cause of diarrhea (*P. shigelloides*). Cells are rod shaped with rounded ends that are motile by polar flagella. Simple chemoorganotrophs (O.J.-II) use ammonium salts and glucose as the sole sources of nitrogen and carbon respectively. Most strains are sensitive to the vibriostatic agent 0/129, are oxidase and catalase positive. Species are found in a wide variety of fish and aquatic animals.

Family C, Section 5, *Pasteurellaceae*

This family includes many parasitic species in vertebrates, and, in particular, the mammals and birds. All species are facultatively anaerobic gRAM negative bacteria that are non-motile and oxidase positive. They are pleomorphic, have coccoid to rod-shaped cells that may swell and also form into filaments. They are all O.J.-III chemoorganotrophs requiring complex media for growth. Some strains are fastidious requiring in addition to several of the amino acids, vitamins, ß-nicotinamide, adenine nucleotides, haematin and/or protoporphyrins. Acids are produced from various carbohydrates, sugar alcohols and glycosides. Nitrate is reduced to nitrite and the catalase and phosphatase tests are commonly positive.

Pasteurella

Species in this genus are parasitic on the mucous membranes of the upper respiratory tract and digestive tracts of mammals and birds but rarely in humans. *P. multocida* is highly pathogenic to mice and rabbits with fatal septicemias being caused by the injection of as few as 1 to 10 cells. The cells are ovoid to rod shaped cells with bipolar staining common. All are non-motile, catalase positive with most strains oxidase positive. Hydrogen sulfide is produced from cysteine, and ornithine decarboxylase is commonly positive; the methyl red and Voges-Proskauer tests are usually negative. Species differentiation is given in Table 3.15.

Haemophilus

In this genus, the species are obligate parasites on the mucous membranes of humans and a variety of animal species. Encapsulated forms of *H. influenzae* are the commonest cause of bacterial meningitis in children and are occasionally associated with acute epiglotitis, osteomyelitis and joint infections. Cells are usually very small to medium sized coccobacilli or rods that sometimes form into threads with a marked degree of pleomorphism. O.J.-IIIf, all strains require preformed factors from blood in order to grow. Of major importance are the X (protoporphyrin IX) and/

or the V (nicotinamide and adenine dinucleotide) factors. Growth is best achieved in very complex media which includes these factors. Denitrification is sometimes complete (to molecular nitrogen). Catalase and oxidase reactions vary between the strains. Carbohydrates are degraded fermentatively yielding acetic, lactic, and succinic acids.

Table 3.15

Species Differentiation of *Pasteurella*

	Beta haem	Growth Mac	Ind	Ure	Cbh	Mant
multocida	-	-	+	-	-	+
pneumotropica	-	-	+	+	-	-
haemolytica	+	+	-	-	-	+
ureae	-	-	-	+	-	+
aerogenes	-	+	-	+	+	-
gallinarum	-	-	-	-	-	-

Beta haem, beta haemolysis; Growth Mac, growth on MacConkey agar; Ind, indole: Ure, urease; Cbh, carbohydrates fermentation generates gas; Mant, mannitol ferments to gas.

Actinobacillus

Members of this genus are generally considered to be opportunistic pathogens. Species have been linked to wooden tongue in cattle (*A. lignieresii*), septicemia in horses (*A. equuli*) and acute septicemia in hogs (*A suis*). Cells are commonly pleomorphic short bacillary or coccobacillary forms that sometimes when the coccoid elements lie in association with the rods give a "Morse code" appearance. Glucose stimulates EPS (slime) production. This EPS may stain faintly pink with the Gram stain. Species are differentiated in Table 3.16. It is readily cultured on the enriched media used to isolate animal pathogens.

Table 3.16

Species Differentiation in *Actinobacillus*

	Hesc	Asuc	Asal	Atre	Alac	Haem
lignieresii	+	-	-	+	-	-
equuli	+	-	+	+	±	-
suis	+	+	+	+	+	+
capsulatus	+	+	+	+	-	+

Hesc,esculin hydrolysis; Asuc, acid from sucrose; Asal, acid from salicin; Atre, acid from trehalose; Alac, acid from lactose; Haem, haemolysis.

Other Genera

These genera have the common characteristics with Section 5 bacteria but do not conveniently fit into any of the three families within the recognized genera described above. An outline of some of the other genera is given in Table 3.17. This is limited to those genera found over a variety of environments or of specific importance. Note that the *Chromobacterium* is described with *Janthinobacterium* in Table 2.19. Another genus of particular interest is *Zymomonas*. This genus has the unique feature (in Section 5) of being able to ferment 1 mol of glucose or fructose to 2 mol of ethanol, 2 mol of CO_2 and some lactic acid. Strains can tolerate 5% ethanol. Cells are rod shaped cells with rounded ends but cells may also be may be ellipsoid. Usually, cells occur in a paired arrangement. Occasionally motile, they possess one to four polar flagella. Strains are all O.J.-III chemoorganotrophs requiring a mixture of amino acids in the culture medium for growth. All strains require biotin and pantothenate. Nitrates are not reduced and oxidase is negative. They are commonly found as a spoiler in fermented products such as beers, wines and perries and are also used as an agent in the fermentation of some plant and honey products.

Table 3.17
Differentiation of Other Genera in Section 5

	gRv	Agl	Pwba	Hpath	Vio	Mot
Zymomonas	-	+	-	-	-	-
Chromobacterium	-	+	-	+	+	+
Cardiobacterium	+	+	+	+	-	-
Eikenella	-	-	+	+	-	-
Streptobacillus	-	+	+	+	-	-

gRv, gRAM variable; Agl, acid from glucose; Pwba, pathogen of warm blooded animals; Hpath, human pathogen; Vio, violacein; Mot, motility.

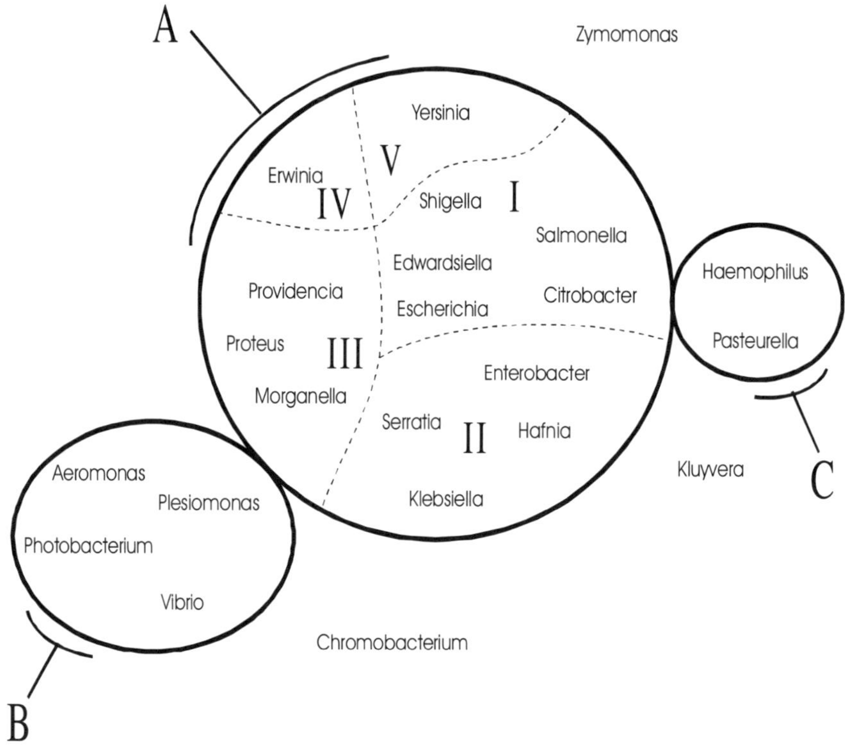

Figure 3.1 Differentiation of the major families in Section 5 and the five tribes that form the *Enterobacteraceae* (family A). These tribes are bordered with dotted lines and are designated as Tribes I, II, III, IV and V. The other two families are designated by the appropriate capital letter. For the *Vibrionacaea* family, the letter used is for B, while for the *Pasteurellaceae* family, the letter C is used.

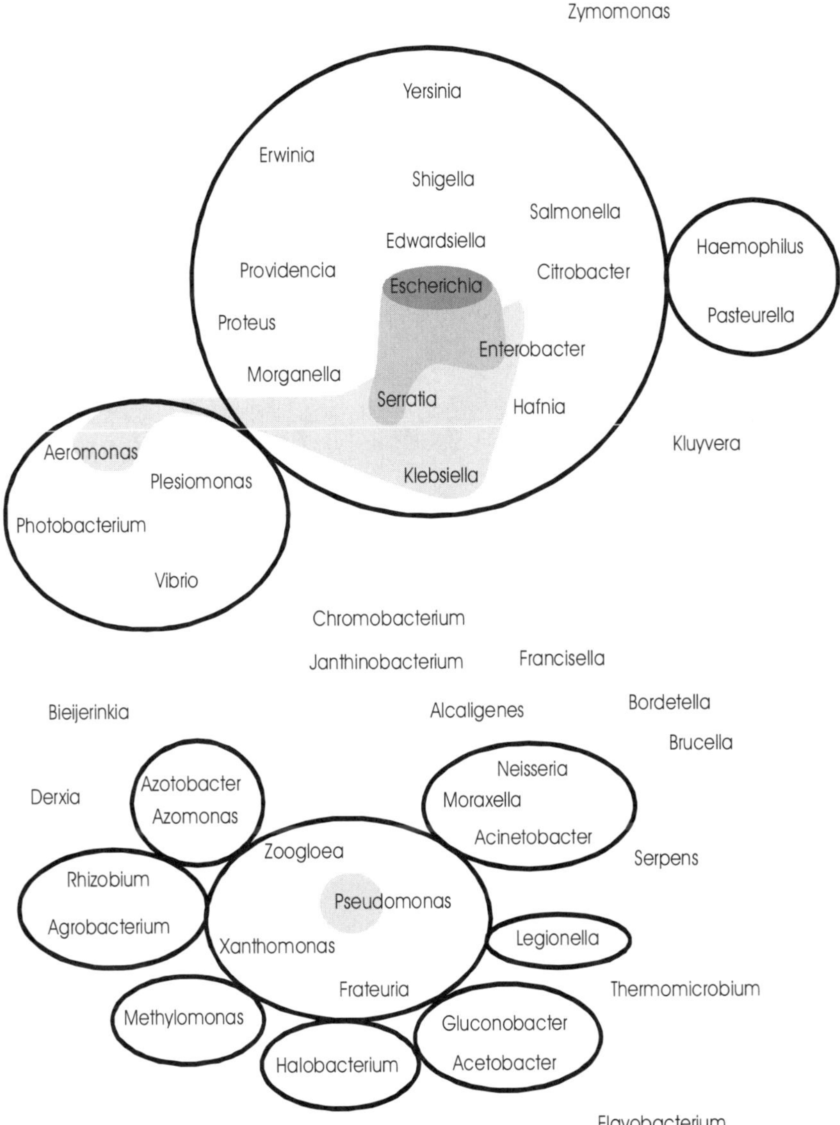

Figure 3.2 Domains of the various groups of coliform genera in the Sections 4 and 5 parts of the atlas. These are differentiated into: *Escherichia coli* (EC, 44.5°C shown as a densely shaded zone); Fecal Coliforms (FC, 35°C, medium shaded); and the Total Coliforms (TC, light shade). See also Table 3.6.

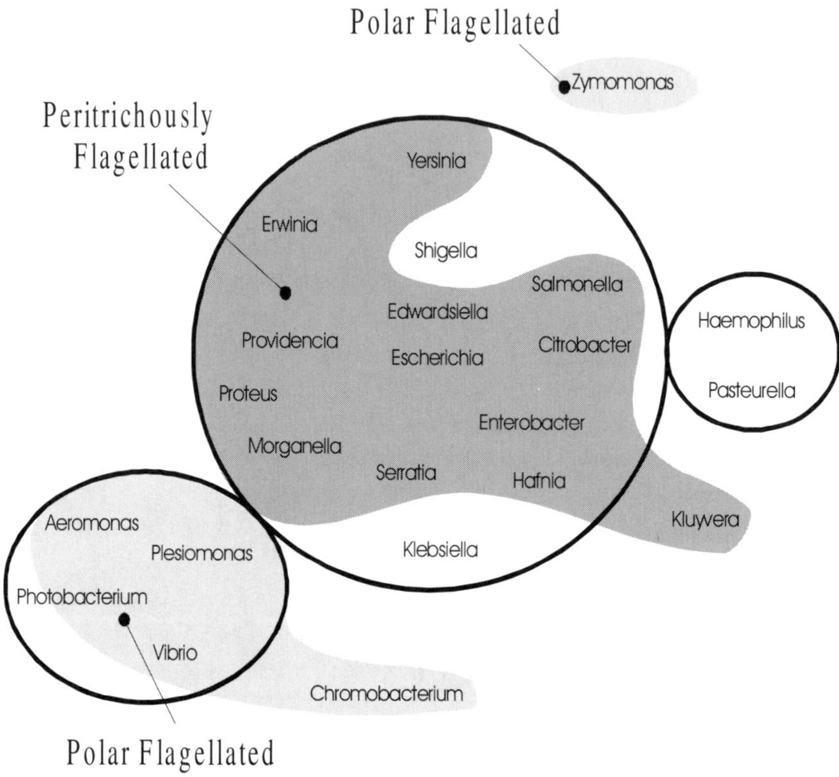

Figure 3.3 Diagram illustrating genera in Section 5 that have peritrichous flagella (dense shade), and polar flagella (light shade). The remaining genera do not possess flagella but *Klebsiella* may possess fimbrae.

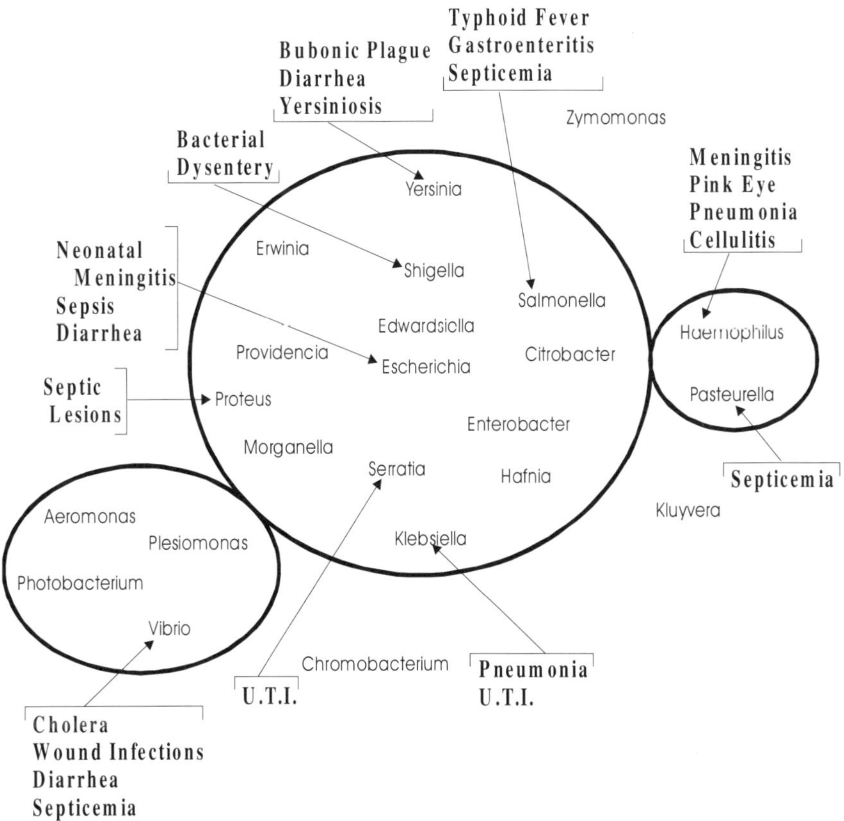

Figure 3.4 Diseases in humans caused by members of Section 5. A separate bracket is displayed for each genus known to be pathogenic to humans. An arrow with a continuous line connects the box to the name of the genus responsible for those diseases. Note that the connections established relate to only some species in the genus and not, necessarily, all species. U.T.I. refers to urinary tract infections.

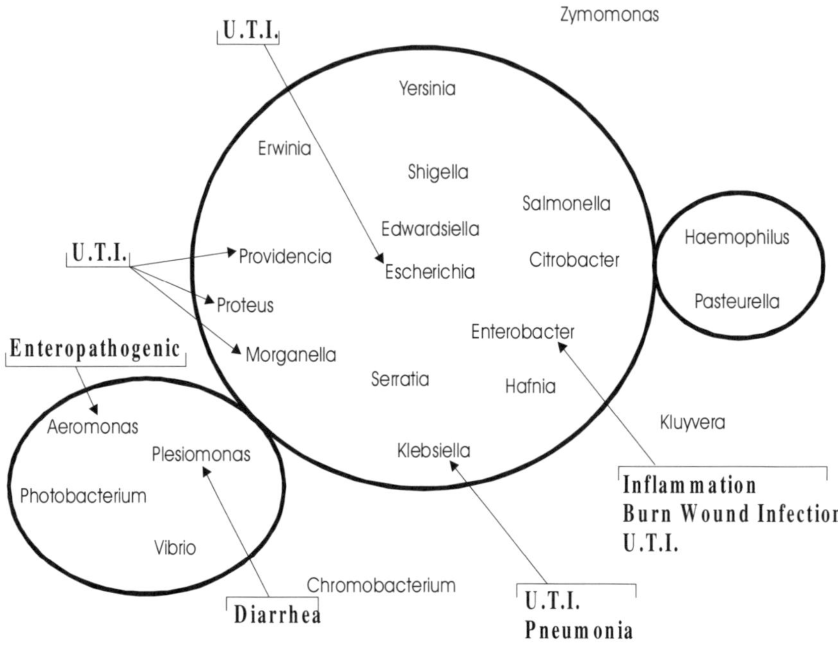

Figure 3.5 Nosocomial diseases in humans caused by some members of the indicated genera of Section 5 bacteria. A separate bracket is displayed for each genus known to be pathogenic to humans. U.T.I. refers to urinary tract infections.

4

gRAM Negative Anaerobic Bacteria

This chapter addresses the bulk of the gRAM negative bacteria that are strictly anaerobic. There are four major sections of bacteria involved in this chapter by section and each section is described below. Major differentiation is by cell shape (Figure 4.1), catalase (Figure 4.2) and flagella patterns (Figure 4.3). Some of the anaerobic bacteria cause diseases in humans (Figure 4.4) The four sections of anaerobic bacteria include:

- Section 6, Anaerobic gRAM negative straight, curved and helical rods. Most strains generate at least a small amount of organic acid during metabolism.
- Section 7, includes the gRAM negative anaerobic bacteria that can reduce either sulfate or sulfur with the generation of hydrogen sulfide. This section is commonly referred to as the sulfate-reducing bacteria (SRB).
- Section 8, includes the gRAM negative strictly anaerobic cocci. They are often found in the gastro-enteric tract of warm blooded animals.
- Section 18, includes the anaerobic anoxygenic phototrophic bacteria. These bacteria are all capable of photosynthesis under reductive conditions (see Figure 4.5).

Section 6, Anaerobic gRAM Negative Straight, Curved, and Helical Rods

Section 6 bacteria are most commonly found in the anaerobic regions of the intestinal tract of warm blooded animals most commonly and, in particular, the ruminants. All in this section are strictly anaerobic gRAM negative rods that are sometimes straight, curved or helical in shape. All are chemoorganotrophs metabolizing at least either carbohydrates, peptones, or metabolic intermediates. Most strains generate at least a small amount of acid products from metabolic activity. There are four genera described below and differentiated in Table 4.1.

Bacteroides

Species in this genus are commonly isolated from a variety of anaerobic environments. For example, *B. fragilis* is commonly isolated from various clinical conditions including appendicitis, periodontitis, rectal abscesses, post-surgical wounds and lesions in the urogenital tract and occasionally isolated from the mouth and vagina. Species are also commonly isolated from human faeces. Other species are isolated from a wide variety of warm blooded animals such as the rumen contents of cattle and sheep. Strains are normally non-motile gRAM negative rods that are strictly anaerobic. Where strains are motile, it is by peritrichous flagella. A wide range of carbohydrates, peptones or metabolic intermediates are metabolized to yield products including succinate, lactate, formate or propionate sometimes with short chained alcohols.

Table 4.1
Differentiation of Genera in Section 6

	Mot	Cell	Substrate	Hab	AcP
Bacteroides	-/+	R	C,P	R.HF	Var
Butyrivibrio	+	CR	C	R.F	B
Leptotrichia	-	RC	C	HO	L
Succinomonas	+	SR,CB	C	BR	S,A

Mot, motile; Cell, cell shape; R, rod shaped; CR, curved rod; RC, rod sometimes in chains; SR, short rod; CB, cocco-bacillary; C, carbohydrate; P, peptone; Hab, common habitat; R, rumen; HF, human faeces; F, feces; HO, human oral cavity; BR, bovine rumen; AcP, acid products; Var, variety of products; B, butyrate; L, lactate; S, succinate; A, acetate.

Butyrivibrio

Species occur commonly in the rumen and sometimes in the feces of humans, rabbits and horses. Cells are usually curved rods that may be helical. All are motile with polar or subpolar flagella. Primary substrate for metabolism are carbohydrates. Where glucose or maltose are fermented, butyrate is a major acid product. Some conditions will revert fermentation from butyrate production to lactic acid.

Leptotrichia

Members of this genus are commonly found in the oral cavity of man and have been associated with dental caries. Cells are straight or slightly curved rods with one or both ends pointed or rounded. Arrangements are in pairs, chains or filaments. While anaerobic, isolated strains will adapt to aerobic culture when under elevated CO_2 concentrations. Carbohydrates are fermented with the production of acid but no gas. The major acid produced is lactic. Acetic and succinic acids may also be produced in trace amounts.

Succinomonas

Species are commonly found in the bovine rumen. The cells are short rods or cocco-bacilli and are normally motile by a single polar flagellum. Carbohydrates such as glucose, maltose, dextrin and starch can all serve as the single source of energy with succinate and acetate being the major products when glucose is fermented. Gas is not produced.

Section 7, Dissimilatory Sulfate and Sulfur-Reducing Bacteria

Known commonly as the SRB (bacteria), this section of bacteria is responsible for many of the problems related to: (1) electrolytic corrosion due to the generation of hydrogen sulfide, (2) presence of "rotten" egg odors as a result of the releases of hydrogen sulfide to the air, and (3) production of black slimes that can plug (clog) porous media and surfaces. The blackness is generated primarily by iron sulfides produced by the hydrogen sulfide reacting with iron salts. It should be remembered that black compounds are also produced when some of the iron carbonates are biologically created.

The SRB are formed by a morphologically diverse group of gRAM negative bacteria that are united by the common ability to utilize an inorganic sulfur compound as the electron acceptor. These compounds are reduced to H_2S which can then trigger such events as the electrolytic corrosive processes in many engineered situations (microbially induced corrosion).

There are two major sulfur substrates the SRB can use. These are sulfate and elemental sulfur. Most strains utilize acetate, lactate and other simple organic fatty acids and have only a limited metabolic ability with carbohydrates that are rarely degraded. Nitrate is reduced. Some strains have been reported to fixate molecular nitrogen. These bacteria are commonly found in anaerobic muds and sediments in freshwater, marine and biofouled environments. Strains are also recovered from the gastrointestinal tract where there have been reports that the SRB are linked to ulcerative colitis.

The SRB are very often found within consortia of other (often aerobic) bacteria and are usually found in the deeper strata of biofilms under reductive environmental conditions. Consequently, SRB can be found in environments where oxygen is present even though they are sensitive to the presence of oxygen. It is the aerobic members of the biofilms that "coat" the SRB and protect them from direct exposure to oxygen. This also applies to the presence of SRB in biocolloidal suspended particles in water containing oxygen. There are three major genera of SRB that are described below.

Desulfovibrio

This genus is the most common of the SRB and is most often associated with corrosion and is readily detected using the SRB-BART™. Cell shape varies from a curved to a sigmoid, spirilloid or occasionally straight rod. They are motile by a single or lophotrichous flagellum. Sulfate and other inorganic oxidized sulfur compounds can act as the terminal electron acceptor that is reduced to H_2S.

To culture this genus, the selective media requires reducing agents to allow growth to occur. Some strains are moderately halophilic which means they are able to grow with salt concentrations in the water exceeding 8%. Most useable organic compounds such as lactate are oxidized to acetate and some strains can use H_2 as the energy source and can assimilate acetate and CO_2, or yeast extract, as carbon sources. Nitrates are sometimes reduced while molecular nitrogen are fixed by other strains. This genus is commonly found in anaerobic muds and sediments in fresh and brackish waters, in the intestines of animals, manure and feces, and in biofouled situations where there has been a complexed biofilm (e.g., slimes, tubercles, nodules) formation. It is the other bacteria in the biofilm consortium (particularly *Pseudomonas*) that causes the reducing environment within which the SRB can thrive. Control strategies aimed against the SRB are often impaired by the protective ability of these "consortial overcoats."

Desulfuromonas

This genus is the "twin" of *Desulfovibrio* except that this genus reduces elemental sulfur and not sulfate. This genus becomes a problem, therefore, when there is free elemental sulfur in the environment. Cells are straight or slightly curved rods and/or elongated ovoid rods commonly motile by a single polar flagellum. All species are strictly anaerobic and these bacteria require elemental sulfur as the terminal electron acceptor which becomes reduced to H_2S (dissimilatory sulfur reduction). L-malate or fumurate may be fermented to acetate and succinate. When grown on agar media, colonies may be translucent to opaque or a light peach to pink color. They can utilize acetate and other simple organic compounds which are completely oxidized to CO_2.

Desulfotomaculum

This forms a "twin" genus to *Desulfovibrio* but this time the difference is that the strains in this genus generate endospores. These endospores are resistant spores that are generated within the cell and are formed by the cell contents. These endospores develop as terminal or subterminal oval to round forms which may, or may not, swell the sporangium. It is, therefore, not a method of reproduction but a method for survival. All strains are strictly anaerobic straight or slightly curved rods which may be motile by either polar or peritrichous flagella. Sulfates, sulfides and other reducible inorganic sulfur compounds can be used as the terminal electron acceptor. Reducible sulfur compounds in a complex medium containing organic growth factors are necessary for the growth of these organisms (O.J.-III). Species are commonly found in soils, fresh water, geothermal regions, the intestines of insects and ruminants as well as in the stratal waters of oil fields.

Section 8, Anaerobic gRAM Negative Cocci

This group is formed by the strictly anaerobic cocci that are parasites of warm blooded animals such as humans, ruminants and rodents where they are commonly found in the alimentary tract. They are routinely catalase negative and characteristically the cells have a paired arrangement; adjacent sides may be flattened. Single cells, masses or chains sometimes occur. O.J.-IIIf, these bacteria require complex media for growth with gas a frequent product of metabolism. Carbohydrates may or may not be utilized. There is a single family with three genera separated (Table 4.2) primarily by their metabolic ability.

Table 4.2

Differentiation of Genera in Section 8

	Cell	CarbF	NO_3	H_2S	AC
Veillonella	C	-	+	+	-
Acidamonococcus	OKD	-	-	-	+
Megasphaera	C	+	-	+	-

Cell, cell shape; C, coccoid; OKD, ovoid and kidney shaped; CarbF, carbohydrates fermented; NO_3, nitrate reduced; H_2S, hydrogen sulfide produced; AC, amino acids can be used as sole energy source.

Veillonella

Species belonging to this genus are commonly parasitic in the mouth, intestinal and respiratory tracts of humans. Coccal cells appear commonly as diplococci or short chains. While catalase negative, an atypical catalase is present in some species. Pyruvate and lactate are fermented but carbohydrates and polyols are not. When lactate is fermented, CO_2 and H_2 along with acetate and propionate are produced. CO_2 is necessary in enhanced concentrations for growth.

Acidaminococcus

Species in this genus are commonly isolated from the intestines of humans and pigs. These bacteria are also often found in feces and in putrid lung abscesses and closed abdominal abscesses. Cells are diplococcal and are commonly oval or kidney shaped. O.J.-III, all are able to metabolize amino acids (especially fumerate) as the main source of energy. Pyruvate and lactate cannot be used but some 40% of the strains can catabolize glucose. Where amino acids are used, the products include acetic acids and butyrate in a molar ratio of 2:1; CO_2 may also be formed but H_2 is not produced.

Megasphaera

Species of *Megasphaera* are found in the rumen of sheep as well as in the feces and intestines of humans. They are large cocci that are commonly in pairs but sometimes also in chains. They are able to ferment lactose with a wide variety of products including acetate, propionate, and four carbon straight and branched chain fatty acids, valerate and copious amounts of carbon dioxide and a little amount of hydrogen. When glucose is fermented, the dominant product is caproate.

Section 18, Anoxygenic Phototrophic Bacteria

It is perhaps hard to conceptualize bacteria that are anaerobic growing with very little light and yet being able to photosynthesize using the longer wavelengths of light that "trickle" down to those depths. Unlike the forms of photosynthesis seen in algae and plants that generate oxygen as a product, Section 18 phototrophic bacteria do not produce oxygen, hence the term anoxygenic. This oxygen is derived from water but Section 18 employs reduced molecules such as H_2S, sulfur, hydrogen, and organic matter as their electron source instead of using water. These bacteria are, therefore, bound by their common ability to photosynthesize without the generation of oxygen (anoxygenic photosynthesis).

Instead of oxygen being produced, some of these bacteria produce elemental sulfur as a terminal product which may then accumulate within or around the cells. These bacteria are known as the "sulfur" bacteria. The second designation depends upon whether the cells are pigmented "green" (because the predominant pigmentation is bacteriochlorophyll) or "purple" if other pigments (primarily xanthophylls) dominate. There is, therefore, a natural division (see Figure 4.5) between purple sulfur bacteria, purple non-sulfur bacteria and green sulfur bacteria. All strains are strictly anaerobic and generally function at very specific reduction-oxidation (redox) potentials under the longer wavelengths of light usually present at only low intensities in turbid waters. Very often these bacteria will cause red or green "blooms" in anaerobic sewage lagoons exposed to sunlight particularly during the summer. Examples are given of genera from each of the three major groupings.

Purple Non-Sulfur Bacteria

Rhodopseudomonas

These are the phototrophic bacteria that do not use hydrogen sulfide nor generate sulfur as a product and yet have a xanthophyl dominance, hence the name "purple non-sulfur" bacteria. These bacteria employ organic molecules as both carbon and electron sources. All are gRAM negative rods that are motile with polar flagella or non-motile. Some cells have gas vacuoles which allow the buoyant density of the cell to float at specific positions in the water column.

Purple Sulfur Bacteria

Chromatium

These bacteria are able to use hydrogen sulfide to generate sulfur and much of the sulfur is deposited internally in the cells as sulfur granules usually within envelopes in the cell membranes. Hydrogen may also serve as an electron donor in this genus. Commonly the "blooms" produced by this genus are red, pink, brown and, more occasionally, purple-violet.

Green Sulfur Bacteria

Chlorobium

Members of *Chlorobium* form the major genus in the green sulfur bacteria. In this genus, the photosynthetic green pigments dominate over the other pigments. These pigments are called chlorosomes and are located within ellipsoidal vesicles that are attached to the cell wall membranes. Sulfur tends to be deposits around the cells in the EPS.

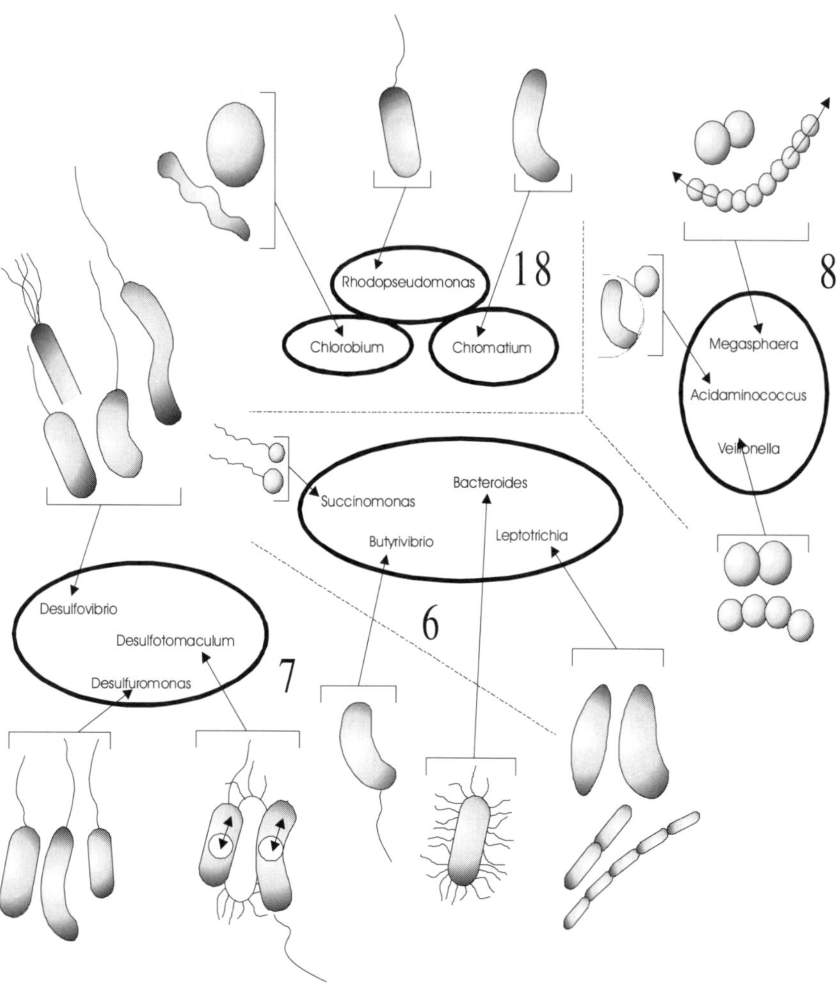

Figure 4.1 Cell shapes observed in the anaerobic bacteria recorded in Sections 6, 7, 8 and 18. These shapes are not in scale to each other but represent the typical form that the cells take during growth. Flagella patterns are also defined in Figure 4.3.

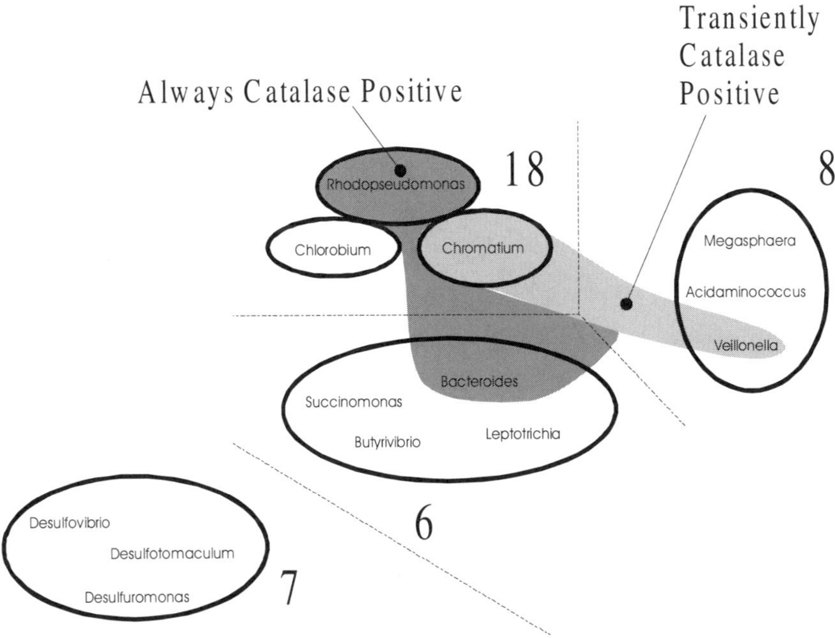

Figure 4.2 Illustration of which genera within the gRAM negative anaerobic bacteria that possess the catalase enzyme system. Genera that are always catalase positive have a dense shade while those that are transiently positive or in which only some species are catalase positive are displayed using a light shade.

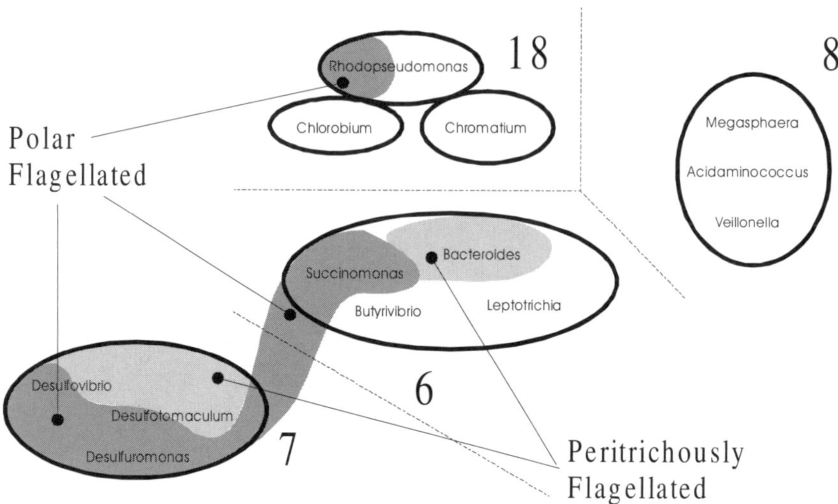

Figure 4.3 Forms of flagella patterns recorded in the anaerobic bacteria recorded in Sections 6, 7, 8 and 18. Polar flagellated genera are outlined using a dense shade while the peritrichously flagellated are defined using a light shade.

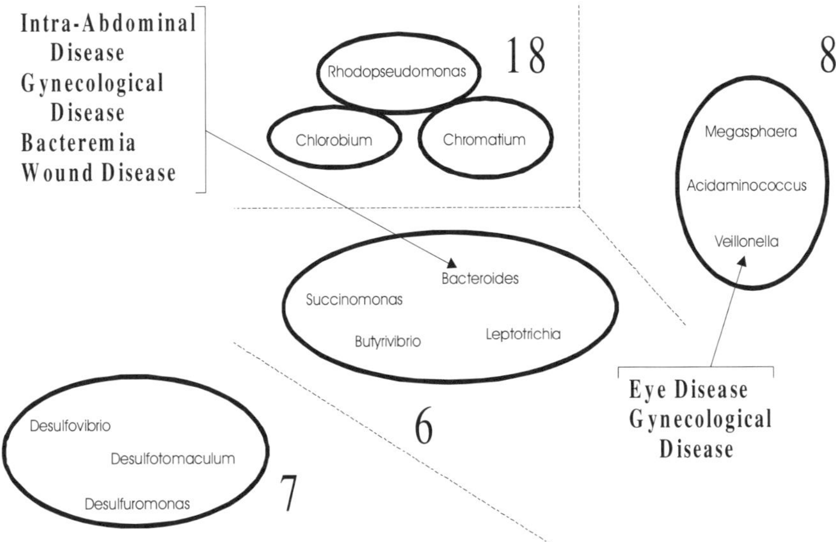

Figure 4.4 Diseases in humans caused by the anaerobic bacteria belonging to Sections 6, 7 and 8. A separate bracket is displayed for each genus known to be pathogenic to humans. An arrow with a continuous line connects the box to the name of the genus responsible for those diseases.

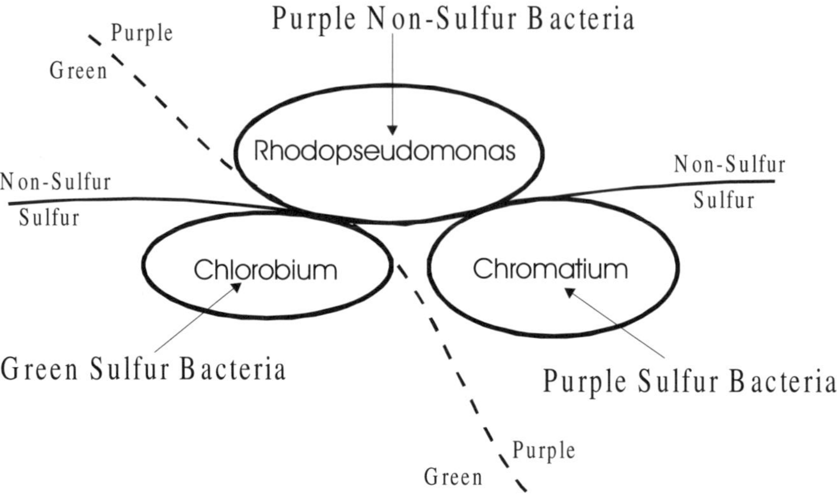

Figure 4.5 Illustration of the primary stages in the differentiation of the phototrophic anaerobic bacteria using the primary form of the pigmentation (green or purple, tracked line) and whether the bacteria accumulate sulfur through the utilization of hydrogen sulfide (sulfur or non-sulfur, continuous line). These lines are used to separate the groups into the three families.

5

gRAM Positive Cocci (Section 12)

This section is formed by a very diverse grouping of the gRAM positive cocci. They are separated into four families (Table 5.1, Figure 5.1) on the basis of aerobicity, the catalase reaction and resistance to gamma radiation. Two families recognized as aerobic cocci are catalase positive. One family is aerobic but catalase negative. The final family is strictly anaerobic and commonly is catalase negative.

Table 5.1

Differentiation of the Families in Section 12

	Aerob	Cata	Gres
A. *Micrococcaceae*	A	+	-
B. *Deinococcaceae*	SA	+	+
C. *Streptococcaceae*	A	-	-
D. *Peptococcaceae*	SAN	-/+	-

Aerob, aerobicity; A, aerobic; SA, strictly aerobic; SAN, strictly anaerobic; Cata, catalase; Gres, resistant to 700 Krad of gamma radiation.

Family A, *Micrococcaceae*

These gRAM positive aerobic cocci are widely distributed in the environment and some are opportunistic pathogens of animals including humans. The cells (see also Figure 5.2) usually divide in more than one plane to form regular or irregular packages or clusters (see also Figure 5.3). Few strains are motile and most are catalase positive. In general, strains can utilize both carbohydrates and/or amino acids as the sources of carbon and energy. There are three genera commonly recognized (Table 5.2) and are easily separated by aerobicity and motility.

Table 5.2

Differentiation of genera in the *Micrococcaceae*

	Aerob	Motility
Micrococcus	SA	-
Staphylococcus	FA	-
Planococcus	SA	+

Aerob, aerobicity; SA, strictly aerobic; FA, facultatively anaerobic.

Micrococcus

Members of this genus occur in a wide variety of aerobic habitats including the mammalian skin, dairy products, soils and water. Species differentiation is given in Table 5.3. While most strains are non-pathogenic, some strains may be opportunistic pathogens. Cells are spherical and are usually in pairs, tetrads or irregular clusters. Always catalase positive and frequently oxidase positive, strains are non-motile. Most produce various yellow carotenoid pigments and most will grow in the presence of 5% NaCl. Glucose can be metabolized with acid production but without gas.

Table 5.3

Species Differentiation in *Micrococcus*

	Pig	Glu	NO_3	NaCl7.5%	Oxi	Urea
luteus	y	-	-	+	+	+/-
lylae	cw	-	-	+	+	-
varians	y	+	+	+	-	+
roseus	p,r,o	+	+	+	-	-
agilis	r	-	-	-	-	-
kristinae	p,o	+	-	+	+	+/-

Pig, pigment; Glu, glucose; NO_3, nitrate reduction; NaCl7.5%, grows in 7.5% NaCl; Oxi, oxidase; Urea, urea hydrolysis; y, yellow; cw, creamy white; p, pink; r, red; o, orange.

Planococcus

Members of this genus are distributed widely in the marine environment and found on clams, prawns and shrimp. Cells are spherical cells occurring singly, in pairs, triads and (less commonly) tetrads and the unique feature is that they are motile by one or two flagella. Protease activity (Figure 5.4) is common with gelatin routinely hydrolysed. Carbohydrates are not attacked. There are two major species that are differentiated (Table 5.4) by salt tolerance, susceptibility to streptomycin and proteolytic ability. They will grow well in salt water supplemented media with strains having a tolerance to NaCl from between 1 to 12%. Colonial growth is commonly dominated by the presence of yellow-orange pigment and a glistening smooth form.

Table 5.4
Species Differentiation in *Planococcus*

	20% NaCl	S. St
citreus	-	S
halophilus	+	R

20%NaCl, growth occurs on 20% salt solution in nutrient agar; S.St, Susceptibility to 5 microg. Streptomycin; S, susceptible; R, resistant.

Staphylococcus

Species of staphylococci are found associated with skin, skin glands and mucous membranes of warm blooded animals (Figure 5.5). They are also found in a variety of animal products such as meat and dairy products. Some strains are opportunistic pathogens (Figure 5.6). Of these, it is the species, *S. aureus,* that has the greatest potential as a pathogen with a range of the major infections in humans ranging from boils (furuncles), carbuncles, impetigo, pneumonia, osteomyelytis, meningitis, mastitis, bacteremia, food poisoning (via an entero toxin), urogenital infection and toxic shock syndrome.

Table 5.5
Species Differentiation in *Staphylococcus*

	Coag	Asuc	Atre	Urease	Nres	Pigm
aureus	+	+	+	+	-	+
epidermidis	-	+	-	+	-	-
warneri	-	+	+	+	-	+/-
haemolyticus	-	+	+	-	-	+/-
hominis	-	+/-	-/+	-	-	+/-
saprophyticus	-	+/-	+	-	-	-

Coag, coagulase; Asuc, sucrose fermented to acid; Atre, trehalose fermented to acid; Urease, urease present; Nres, resistant to MIC >1.6 microg/ml of novobiocin; Pigm, carotenoid pigment.

Characteristically, the cells in this genus divide in more than one plane to form irregular clusters but singles, pairs and tetrads are also found. Strains are usually catalase positive and non-motile facultative anaerobes. Species are differentiated in Table 5.5. Aerobic growth is, however, much more copious than under anaerobic conditions. Most strains will grow well on media with 10% NaCl. There is a variable fastidiousness (O.J.-II and O.J.-III). Most require some of the amino acids for growth while others can grow with $(NH_4)_2SO_4$. While strains are not susceptible to lysis by lysozyme, they are susceptible to lysis by lysostaphin.

Family B, *Deinococcus*

This family is unique because of the tremendous resistance of members of this family to gamma radiation. It could be argued that this would be the last genus left alive on this planet after a very severe release of gamma radiation. The cells are constructed with very unusual cell wall structures. The cell walls show a complex electron microscopic profile with several layers of distinctive components. The lipid composition includes palmitoleate (16:1) and an unusual spectrum of polar lipids. There is only one genus, *Deinococcus,* in this family.

Deinoncoccus

This genus includes all of the strains of cocci that are able to resist radiation dosages of up to 500 Jm^{-2} for UV-radiation and 700 Krad for gamma radiation. They were first recovered from gamma irradiated foodstuffs where they had caused spoilage in the sealed product. Cells are spherical cocci that appear to be large, partly because of the outer membranes and the presence of several distinct cell wall layers. All are catalase positive and have a strictly respiratory metabolism. A range of carbohydrates are metabolized but only a few (such as glucose) will produce acid (no gas). Carotenoids are present in all species and colonies are usually pink to brick red. All grow well on tryptone glucose yeast extract medium (TGYM). There are four major species (Table 5.6).

Table 5.6

Species Differentiation in *Deinococcus*

	5% NaCl	Agl	Afr	NO_3	Pig
radiodurans	+++/-	-	+++/-	-	r
radiophilus	+	-	-	-	o-r
proteolyticus	-	+	-	-	o-r
radiopugnans	-	-	+	-	o-r

5%NaCl, growth; Agl, acid from glucose; Afr, acid from fructose; NO_3, nitrate reduced; Pig, carotenoid pigmentation; o, orange; r, red.

Family C, *Steptococcaceae*

These were recognized in the eighth edition of Bergey's Manual as a distinct family but relegated to "other genera" status in the ninth. These bacteria are gRAM positive, catalase positive cocci, many of which divide in a single or multiple plane of division. Commonly fastidious O.J.-III with complex nutritional requirements. There are four major genera that are differentiated in Table 5.7 on the basis of aerobicity, cell arrangement and fermentation substrates and products.

Table 5.7
Differentiation of Genera in Family C, *Streptococcaceae*

	Aerob	Cellar	CfermP
Streptococcus	FA	Ch	L
Leuconostoc	FA	Pr,Ch	L,A/E
Pediococcus	FA	Pl,Tr	L
Aerococcus	MA	Tr	W

Aerob, aerobicity; FA, facultatively anaerobic; MA, microaerophillic; Cellar, cell arrangement; Ch, chains; Pr, pairs; Pl, plainar; Tr, tetrad; CfermP, carbohydrate fermentation acidic products; L, lactic; A/E, acetic and/or ethanol; W, weak.

Streptococcus

This genus includes many species that are commensals or parasites on various animals including humans. Others are saprophytic and associated with the spoilage or production of foods. Species of particular interest include: *S. pyogenes* (streptococcal sore throat, acute glomerulonephritis, rheumatic fever), *S. pneumoniae* (lobar pneumoniae), *S. faecalis* (also known as *Enterococcus faecalis*, opportunistic pathogen causing urinary tract infections and commonly found in the intestines of many warm blooded animals including humans), and *S. lactis* (used as a starter bacteria for cheese production).

There are five major species separated in Table 5.8 and four species groups are separated based upon their ability to grow at 10° and 45°C. Each of the first four groups are differentiated in Table 5.9 (pyogenic streptococci), Table 5.10 (oral streptococci), Table 5.11 (enterococci/fecal streptococci), and Table 5.12 (lactic streptococci).

Cells are moderately sized spherical to ovoid cells which are usually seen in pairs or chains when grown in liquid media. In some cases, these chains may be very long during early culture. Carbohydrates are fermented with production (mainly) of some form of lactic acid but no gas.

Table 5.8

Differentiation of the Major Species Groups of *Streptococcus*

Group	G10°	G45°	6.5%Ncl	pH9.6	Aerob
pyogenic	-/+	-	-/+	-	FA
oral	-	-/+	-/+	-	FA
enterococci	+	+	+	+	FA
lactic	+	-	-	-	FA
anaerobic	-	+	-	-	AN

G10°, growth at 10°; G45°, growth at 45°; 6.5%Ncl, growth in 6.5% NaCl; pH9.6, growth at a pH of 9.6; Aerob, aerobicity; FA, facultative anaerobe; AN, strict anaerobe.

Table 5.9

Differentiation of Species in the Pyogenic Group of *Streptococcus*

	Haem	HipH	AfCf
pyogenes	-	-	L
agalactiae	-/+	+	L,R
equi	-	-	-
dysgalactiae	+	-	L,R
pneumoniae	+	-	L,R

Haem, alpha haemolysis; HipH, hippurate hydrolysis; AfCf, acid from carbohydrate fermentation; L, lactic; R, ribose.

Table 5.10

Differentiation of Species in the Oral Group of *Streptococcus*

	AfCf	Harg	Hesc
salivarious	R	-	+
sanguis	R	+	+/-
mutans	M,R,S	-	+
rattus	M,R,S	+	+
mitis	-	+	+

AfCf, acid fermented from carbohydrates; R, ribose; M, mannitol; S, sorbitol; Harg, hydrolysis of arginine; Hesc, hydrolysis of esculin.

Table 5.11
Differentiation of Species in the Enterococcal Group of *Streptococcus*

	AfCf	Haem
faecalis	So,M	+
faecium	A, SO, SR	-
avium	A, SR, SO, M	-
gallinarum	A, SO, M	-

AfCf, acid from carbohydrate fermentation; So, sorbitol; M, melezitose; A, arabinose; SR, sorbose; Haem, beta haemolysis.

Table 5.12
Differentiation of Species in the Lactic Group of *Streptococcus*

	10°	45°	40%b	Harg	Hhipp	Hesc
lactis	+	-	+	+/-	+/-	+/-
raffinolactis	+	-	-	-	+/-	-
uberis	+	-	+/-	+	+	+
bovis	-	+/-	+	-	-	+
thermophilus	-	+	-	+/-	-	-

10°, growth at 10°C; 45°, growth at 45°C; 40%b, grows in 40% bile; Harg, hydrolysis of arginine; Hhipp, hydrolysis of hippurate; Hesc, hydrolysis of esculin.

Leuconostoc

Species occur widely in plants, silage and milk and are used in the production of sauerkraut, silage, pickles, cheeses, butter and buttermilk. Cells are cocci that are commonly elliptical or elongated and are commonly arranged in chains or pairs. Fastidious O.J.-III, species often requiring complex growth factors as well as some of the amino acids for growth. Species differentiation is given in Table 5.13. In *Leuconostoc*, growth is dependent upon the presence of a fermentable carbohydrate. Glucose is fermented to D-lactate and ethanol or acetic acid by means of the phosphoketolase pathway. Some species generate large slime formations around colonies in high sugar content media.

Table 5.13
Species Differentiation in *Leuconostoc*

	Aara	Afru	Asuc	Atre	Dex	pH4.8
mesenteroides	+	+	+	+	+	-
dextranicum	-	+	+	+	+	-
cremoris	-	-	-	-	-	-
paramesenteroides	+/-	+	+	+	-	+/-
lactis	-	+	+	-	-	-
oenos	+/-	+	-	+	-	+

Aara, acid from arabinose; Afru, acid from fructose; Asuc, acid from sucrose; Atre, acid from trehalose; Dex, dextran formation; pH4.8, will grow at a pH of 4.8.

Pediococcus

Species belonging to this genus occur widely on plants and fermenting vegetative products. This genus is commonly involved in the spoilage of beer and the production of sauerkraut and silage. Cells are always spherical, never elongated, with division being in alternate planes at right angles to each other. As a result, the cells commonly appear in the tetrad arrangement, but occasional pairs are also seen. Speciation is given in Table 5.14. Fastidious O.J.-III, strains often require nicotinic acid, pantothenic acid and biotin as well as other complex growth factors and amino acids. Glucose is fermented with either DL- or L-(+) lactate as the principal acidic product.

Table 5.14
Species Differentiation in *Pediococcus*

	Growth°C	Acid from	pHGrowth	Cat
damnosus	<35	S	4.2	-
parvulus	35	-	4.2	-
dextrinicus	40	L,S	-	-
acidilactici	50	L	4.2 & 8.5	+
halophilus	35	S	8.5	-

Growth °C, upper growth temperature; S, sucrose; L, lactose; pHGrowth, upper and lower limits of pH for growth when outside the normal range of 6.5 to 8.0; Cat, catalase.

Aerococcus

This is a genus commonly found as an airborne contaminant in hospitals and also as the cause of a fatal disease (gaffkemia) in lobsters. They grow under conditions where there are low oxygen concentrations (microaerophilic). Cells are spherical and, in liquid media, form tetrad arrangements. In semi-solid agar media, colonies tend to form beneath the surface of the agar as a distinct band (layer). Catalase is negative or weakly positive. Growth on solid agar media is generally sparse and beaded with small discrete colonies. Where blood agar is used, colonies will produce a "greening" reaction.

Family D, Anaerobic Group of gRAM Postivie Cocci

All of these bacteria are gRAM positive cocci that are strictly anaerobic fastidious O.J.-III and normally catalase negative. There are four major genera that are differentiated in Table 5.15.

Table 5.15
Differentiation of Genera in Family D, Anaerobic gRAM Positive Cocci

	FerC	FerP	CellA	HfC	HfP
Peptococcus	-	+	C	-	+
Peptostreptococcus	-	+	Ch,T,I	-	-
Rumninococcus	+	-	Dc,Ch	+	-
Sarcina	+	-	S	+	-

FerC, ferments carbohydrates; FerP, ferments peptones; CellA, cell arrangement; C, coccoid; Dc, diplococcal; Ch, chains; T, tetrad; I, irregular clumps; S, sarcinoid (cuboidal packets of eight or more cells); HfC, hydrogen generated from carbohydrates; HfP, hydrogen generated from peptones.

Peptococcus

Species are isolated from various clinical specimens. They are able to metabolize amino acids and peptones producing a range of normal and branched chained fatty acids in combinations from C1 through to C6. Copious hydrogen can be produced from peptone but carbohydrates are not fermented.

Peptostreptococcus

Species are found widely dispersed in clinical specimens and vertebrate and invertebrate tissues. The coccoid cells form into a variety of arrangements including pairs, tetrads, irregular masses and chains. Peptone and amino acids are metabolized with acetic acid as the major product but with isobutyric, butyric, isovaleric or isocaproic acids also being generated as minor products. Culturing is usually performed on enriched media such as blood agar with the surfactant, Tween 80. This will sometimes enhance growth.

Ruminococcus

Members of this genus are commonly found in the gastro-enteric tract of many warm blooded animals with the most common sites of isolation being the rumen, caecum and large bowel. Cells may be either spherical or elongated. Where elongated, the tips of the cells may be pointed. Cells are commonly arranged in chains or diplococci. Carbohydrates can be fermented with acid and gas (H_2 and CO_2) production. Acetate, succinate, lactate and ethanol are among the products commonly produced. Amino acids and peptones are not fermentable. Catalase negative, nitrate is not reduced and ammonium is not produced from amino acids or peptones.

Sarcina

Sarcina contains species that will grow over a wide pH range from 1.0 to 9.8. They are mostly O.J.-III and numerous amino acids and some vitamins are required to achieve growth. Species are found in a wide variety of habitats including soil, mud, grain, faecal material and some clinical specimens (e.g., stomach). Cells are nearly spherical cells and are arranged in cuboid packets of eight or more cells (known as the sarcinoid arrangement). Adjacent sides of the cells in contact with other cells may be flattened. Carbohydrates are commonly fermented with the major products being CO_2, H_2 and either ethanol or butyric acid (depending on the species).

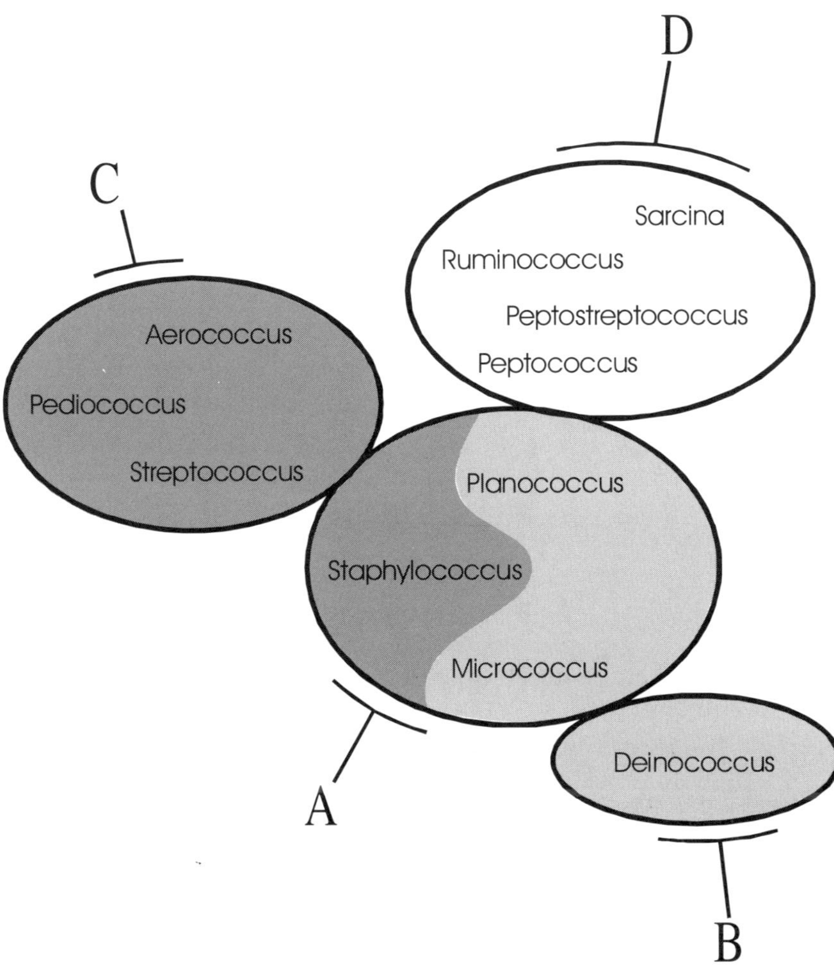

Figure 5.1 Diagram showing the primary differentiation of the gRAM positive cocci in Section 12 based upon their aerobicity. The strictly aerobic (SA, light shade) are separated from the facultatively anaerobic bacteria (FAN, dense shade) and the strictly anaerobic cocci (SAN, unshaded).

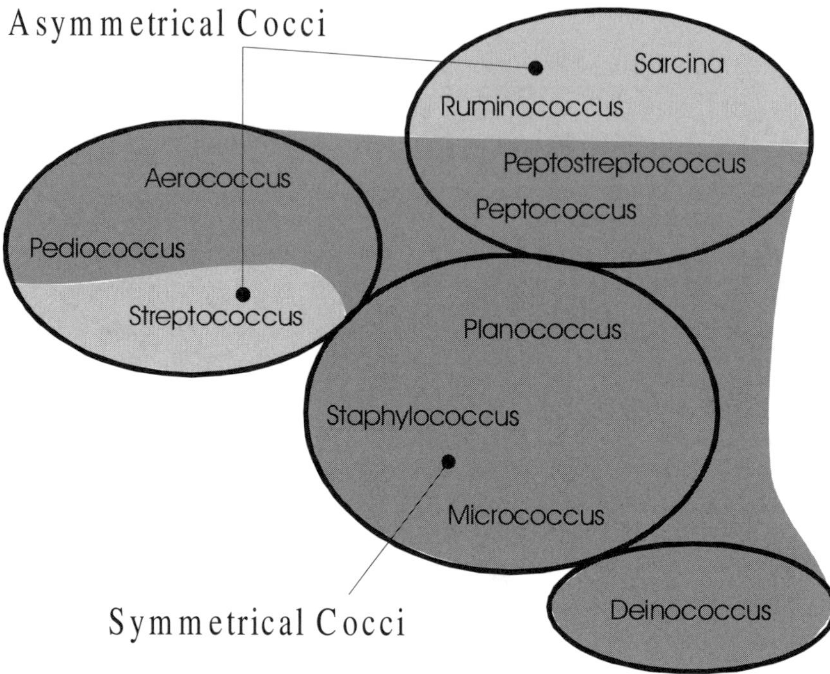

Figure 5.2 This figure illustrates the two forms of coccoid cell shapes that are seen in the Section 12 gRAM positive coccoid bacterial group. There are two primary forms, symmetrical (dense shade) and asymmetrical (light line), which separate the bacterial genera into two groups.

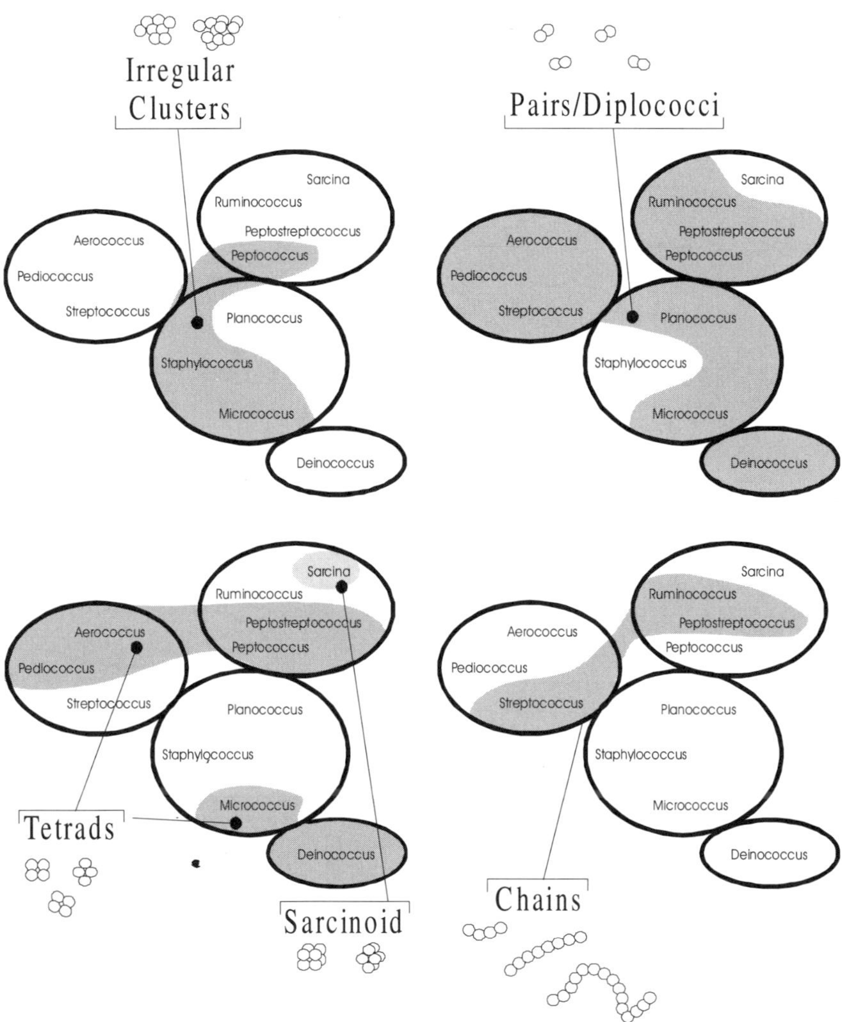

Figure 5.3 Illustration of the four principal cell arrangements found among the bacterial genera in Section 12. Note that several different cell arrangements may, at various times, be seen within the same genus.

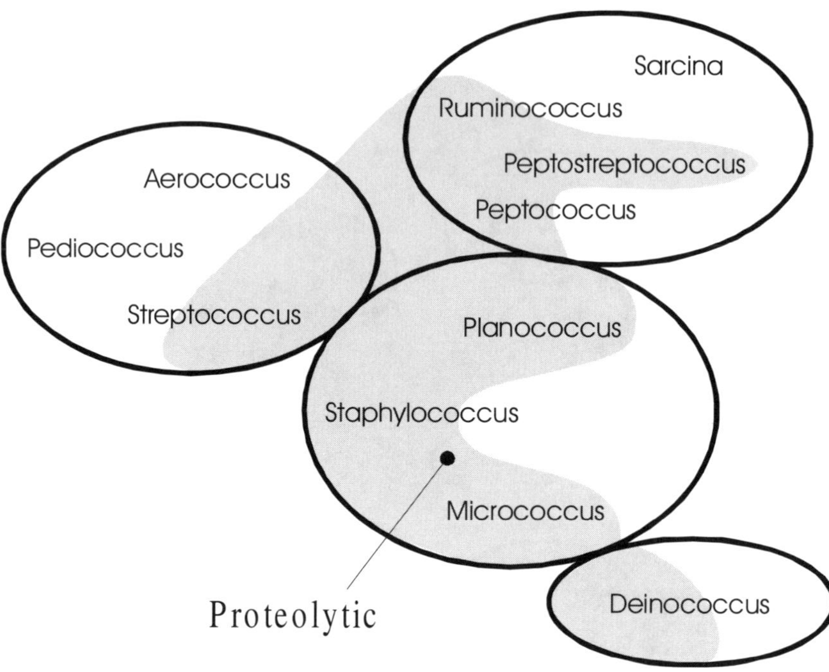

Figure 5.4 Proteolysis (shaded) is a distinctive feature that allows some of the species to be separated during identification in Section 12 gRAM positive cocci.

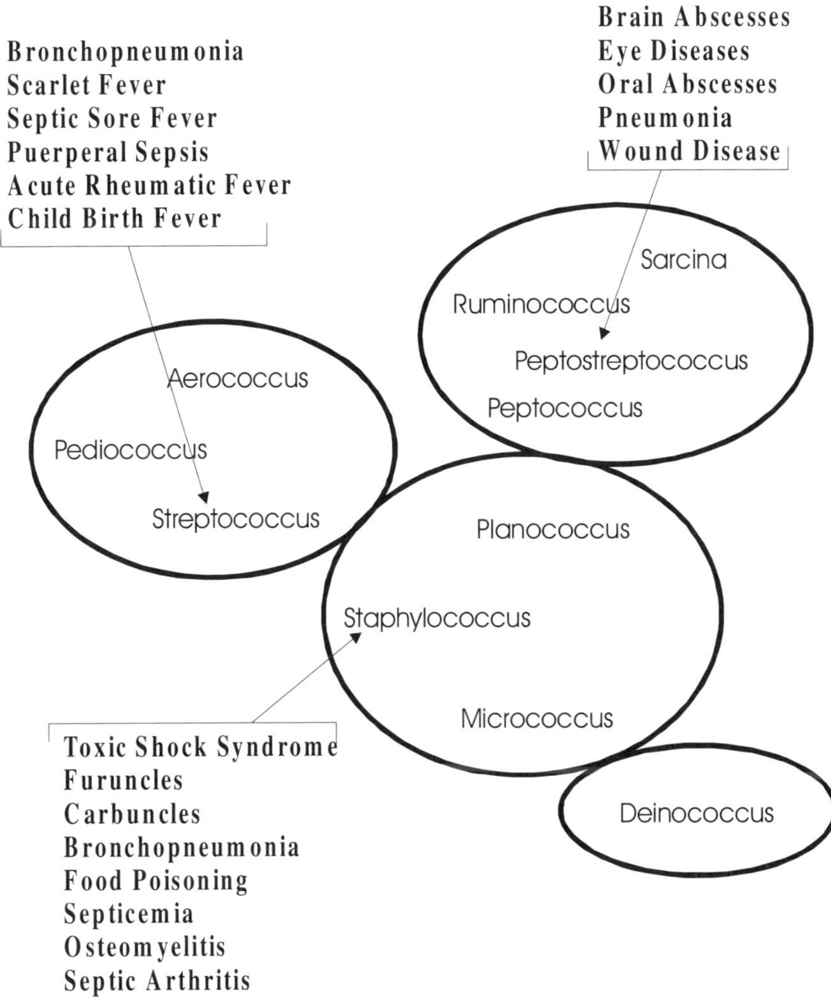

Figure 5.5 Diseases in humans caused by members of Section 12, the gRAM positive cocci. A separate bracket is displayed for each genus known to be pathogenic to humans. An arrow with a continuous line connects the box to the name of the genus responsible for those diseases.

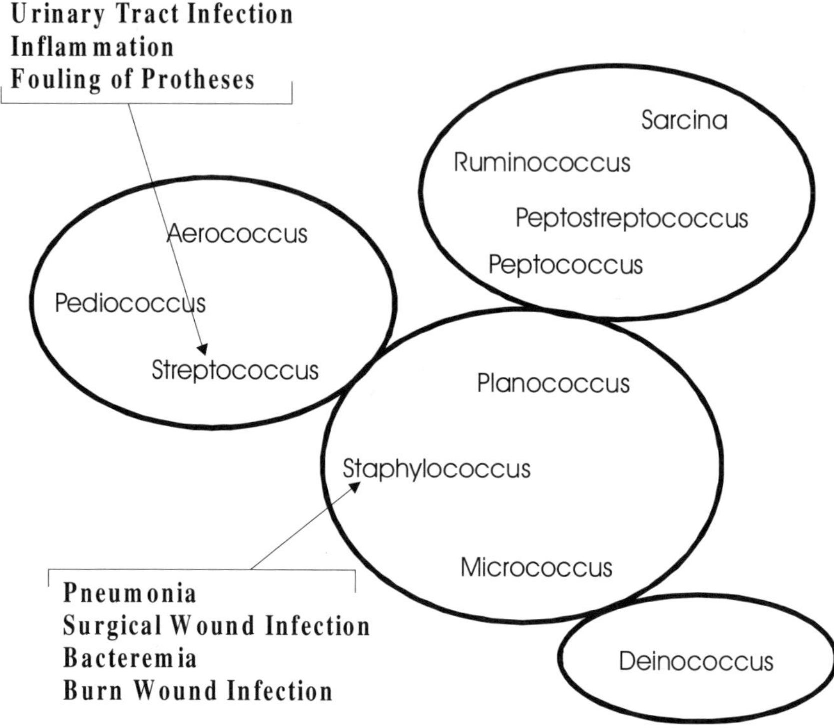

Figure 5.6 Nosocomial diseases in humans caused by members of Section 12, the gRAM positive cocci. A separate bracket is displayed for each genus known to be pathogenic to humans. An arrow with a continuous line connects the box to the name of the genus responsible for those diseases.

6

gRAM Positive Rods

The gRAM positive rods are a group incorporating four sections of bacteria. These bacteria do not produce complex branching web-like masses of filamentous growth (mycelium) or specialized extensions from the cells. They are differentiated in Table 6.1 (Figure 6.1). Separation is by those that have an internal survival cell (endosporogenous, Section 13). Others are non-sporing rod-shaped in a regular form (regular asporogenous rods, Section 14). Aerobicity (Figure 6.2) and cell form (Figure 6.3) further separates those that do not have a regular cell form (irregular asporogenous rods, Section 15). The acid-fast positive rods are mycobacteria (Section 16).

Table 6.1

Differentiation of the Sections in the gRAM Positive Rods

Section	Name	Endospores	Acid-Fast	Cellf
13	Endosporogenous	+	-	R
14	Regular	-	-	R
15	Irregular	-	-	R
16	Mycobacteria	-	+	RM

Cellf, cell form; R, regular; RM, regular with mycelia occurring sometimes as a part of the life cycle.

There is a range of diseases in the human species as illustrated in Figure 6.5.

Section 13, Endospore-Forming gRAM Postive Rods and Cocci

This section of bacteria probably forms the most durable and robust group because they are able to produce endospores. In simple terms, endospores (Figure 6.6) are survival cells formed within the cell by the condensation of the essential cell components into a smaller protected region of the cell (sporangium) generating the spore. The water content is much lower in the spore than the sporangium and the spore is protected by a thick spore wall which is resistant to adverse conditions (e.g., high temperatures, very high or low pH or redox conditions). The endospore has this name because it is inside (endo-) the cell and is easily recognized since it does not take up the gRAM stain but remains clear. The endospore is not a means of reproduction but a means of survival and there are records of these endospores still remaining viable after millennia. The "survivor" is the condensed cell inside the spore. When the environmental conditions are favorable, the spore wall cracks open much like a seed case and the cell swells as it emerges through the crack. There are five genera in this section (Table 6.2) that are differentiated by aerobicity and cell form.

Endospores are most easily categorized using the spore stain. Here, the endospore becomes stained and it is now possible to determine the shape, size and position of the spore (Figure 6.6). This is particularly important in the speciation of *Bacillus* (Table 6.3). Shapes vary from spherical to ovoid, ellipsoid and cylindrical. Size is determined by whether the endospore becomes large enough to press out the side walls of the rod. If it does, it is called swelling the sporangium. Position is judged by the location of the spore along the cell. When the spore is right at the end of the sporangium so much so that no stained cell can be seen beyond the spore, this is know as a terminal location. If some stained cell contents can be seen, then it is sub-terminal. Where the spore appears to be at the "dead center" of the cell, it is centrally located. Any other position may be considered as para-central.

Table 6.2
Differentiation of the Genera in Section 13, Endosporogenous gRAM Positive Rods

	Catalase	Aerobicity	Cellf
Bacillus	+	SA, FA	R
Clostridium	-	SAN	R
Sporolactobacillus	-	MA	R
Sporosarcina	+	A	CT
Oscillospira	-	FA	FD

SA, strictly aerobic; FA, facultatively anaerobic; SAN, strictly anaerobic; MA, microaerophillic; A, aerobic; Cellf, cell form; R, rod; CT, coccoid tetrad; FD; filamentous with disk shaped cells.

Bacillus

Species of *Bacillus* are widely distributed in many natural habitats and relatively few can cause human infections. An exception is *B. anthracis* which causes anthrax in humans and some other warm blooded animals. *Bacillus* species are, however, the scourge of the insect world since there are a range of species that can cause infections in insects including *B. thuringiensis* (generates a parasporal crystal which is toxic to the larvae of *Leptidoptera*), *B. larvae* (American foulbrood of honeybees) and *B. popilliae* (milky disease of the Japanese beetle). Species are identified by the position of the endospore, the Voges-Proskauer reaction, hydrolysis, and acid production from sugars (Table 6.3).

Table 6.3
Species Differentiation in *Bacillus*

	EndoR	EndoS	V.P.	AcidS	HydrO
subtilis	-	-	+	A,X,M	C,G,S
anthracis	-	-	+	-	C,G,S
cereus	-	-	+	-	C,G,S
firmus	-	-	-	-	C,G,S
lentus	-	-	-	A,X,M	C,G,S
marinus	+	-	-	X	C,G
megaterium	+/-	-	-	A,X,M	C,G,S
pumilis	-	-	+	A,X,M	C,G
thuringiensis	-	-	+/-	-	C,G,S
alvei	-	+	+	-	C,G,S
brevis	-	+	-	M	C,G,S
circulans	-	+	-	A,X,M	C,G,S
coagulans	-	+	+	A,X,M	C,G,S
globiformis	+	+	-	-	C,G,S
laterosporus	-	+	-	-	C,G
lentimorbus	-	+	-	-	-
macerans	-	+	-	A,X,M	-
polymyxa	-	+	+	A,X,M	C,G,S
popilliae	-	+	-	-	-
sphaericus	+	+	-	-	C,G
stearothermophilus	-	+	-	A,X,M	C,G,S

EndoR, endospores round; EndoS, endospores swelling rod; V.P., Voges -Proskauer; AcidS, acid from sugars; A, arabinose; X, xylose; M, mannitol; HydrO, hydrolysis of organic compounds; C, casein; G, gelatin; S, starch.

Most strains are catalase positive and aerobic with oxygen being the common terminal electron acceptor. This is replaceable in some species by alternatives (e.g., nitrate). Strains are either aerobic or facultatively anaerobic straight rod shaped bacteria which will sporulate always under aerobic conditions. Strains are commonly motile by peritrichous flagella which may be degenerated as the cells mature. Some species are oxidase positive. *Bacillus* is a very diverse genus in which both O.J.-II and -III strains are found.

Clostridium

Species of *Clostridium* form a very diverse group of strictly anaerobic endosporogenous bacteria that are found in many anaerobic (reductive) environments (Figure 6.8). They are commonly found in reductive soils, sewage, marine sediments, decaying organic material, and in the intestinal tracts of a wide range of animals including humans. Many species are pathogenic in warm blooded animals including humans. Many strains produce copious gas or toxic products as a part of the infectious process. Examples of this include: *C. botulinum* (botulism), *C. perfringens* (gas gangrene, bacteremia), *C. tetani* (tetanus), and *C. septicum* (braxy in sheep; malignant edema in cattle). Other major clostridia species include *C. pasteurianum* (able to fix molecular nitrogen), *C. acetobutyricum* (production of butanol), and *C. cellobioparum* (initial degradation of cellulose). Cells are rod shaped usually staining gRAM positive, particularly during the early stages of growth. When motile, the flagella are peritrichous. Endospores are usually oval to spherical and swell the cell (sporangium). All are catalase negative and the endospores cannot be produced under aerobic conditions. Some strains can, however, tolerate oxygen but will not sporulate.

Species of *Clostridium* are divided into four groups on the basis of the fermentation of glucose to acid and the hydrolysis of gelatin (Table 6.4) and the species are differentiated within those four groups. Group one is given in Table 6.5, group two in Table 6.6, group three in Table 6.7, and group four in Table 6.8.

Table 6.4

Differentiation of Species of *Clostridium* into Four Groups

	Gelatin hydrolysis	Glucose to acid
Group 1	-	+
Group2	+	+
Group 3	-	-
Group 4	+	-

Table 6.5
Species Differentiation in Group One, *Clostridium*

	Mot	Ind	Lip	Acab	Lmilk	Digest
botulinum	+	-	+	-	c/d	+
chauvoei	-	-	-	l,s	cu	-
haemolyticum	+	+	-	-	cd	-
novyi	+	+	-	-	cd/-	+
perfringens	-	-	-	c,l,s	cd	+
roseum	-	-	-	c,l,s	cd	-
septicum	+	-	-	c,l	cd	-
sporogenes	+	-	+	-	d	+

Mot, motile; Ind, Indole; Lip, Lipase; Acab, acid from carbohydrates; c, cellobiose; l, lactose; s, sucrose; Lmilk, litmus milk; c/d, curd or digestion; cu, curd; cd, curd with digestion; d, digestion; Digest, meat digestion.

Table 6.6
Species Differentiation in Group Two, *Clostridium*

	Mot	H_2prod	Esc	Starch
acidurici	+	-	-	-
aminovalericum	+	++++	+	+
kluyveri	+	++	-	-
leptum	-	++++	+	-
polysaccharolyticum	+	+++	+	+
sporosphaeroides	-	++++	-	-
sticklandii	+	+	-	-

Mot, motility; H_2prod, hydrogen production; Esc, esculin hydrolysis; Starch; starch hydrolysis.

Table 6.7
Species Differentiation in Group Three, *Clostridium*

	Mot	Ind	Esc	Str	Nit	Acarb
acetobutylicum	+	-	+	+	-	c
beijerinckii	+	-	+	+	-	c,l,s
butyricum	+	-	+	+	-	c,l,s
cellobioparum	+	-	-	+	-	c
durum	+	-	+	-	-	l,s
fallax	-	-	+	-	-	s
pasteurianum	-	-	-	-	-	l,s
saccarolyticum	-	+	+	+	+	-
thermaceticum	-	-	-	-	+	-
thermo-saccharolyticum	+	-	+	+	-	c,l,s

Mot, motility; Ind, indole; Esc, esculin hydrolysis; Str, starch hydrolysis; Nit, nitrate reduced to nitrite; Acarb, acid from carbohydrates; c, cellobiose; l, lactose; s, sucrose.

Table 6.8
Species Differentiation in Group Four, *Clostridium*

	Mot	Ind	Lec	H_2 prod
ghonii	+	+	+	++
tetani	-/+	+/-	-	++++
botulinum	+	-	-	++++
villosum	-	-	-	++
limosum	+/-	-	+	-
histolyticum	+/-	-	-	++
putrifaciens	-	-	-	-

Mot, motility; Ind, indole; Lec, lecithinase; H_2prod, hydrogen produced.

Sporosarcina

This genus has been reported being found in a wide variety of habitats including soils, salt marshes, in the respiratory tracts of animals and humans, and has also been associated with the degradation of urea, often found at sites where dogs urinate, particularly against trees. Cells are spherical to ovoid and reproduce in a series of perpendicular planes to form tetrads and cuboidal packages of eight or more cells (sarcinoid). Commonly, the cells are motile by one or two randomly placed flagella. All are aerobic with catalase and oxidase reactions normally positive. Moderately fastidious O.J.-III, colonies are often pigmented from a cream color through to a pale yellow or bright orange color.

Oscillospira

This is a genus that occurs commonly in the gastro-enteric tract of herbivorous animals. Cells are large gRAM negative rods or filaments with disk shaped cells that can be motile by numerous lateral flagella under very reductive conditions. Motility is lost upon exposure to air. Endospores are formed longitudinally in the rod shaped cells and almost fill the disk shaped cells. On rare occasions a single cell may bear two endospores.

Sporolactobacillus

This genus was isolated originally from chicken feed but, subsequently, has been found in a variety of soils but in low numbers. Cells are straight rods that occur singly, in pairs and (rarely) in short chains. All strains are motile by sparsely peritrichous long flagella. They are sensitive to high concentrations of oxygen but are microaerophiles that generate only one form of lactic acid in the fermentation of hexoses with the usual product being D-lactic acid. All strains are catalase negative, moderately fastidious O.J.-III and can be grown on a peptone glucose yeast extract medium.

Section 14, Regular Non-Sporing gRAM Positive Rods

This section is easily recognized since it is comprised of gRAM positive rods that have a regular symmetrical (mirror imaged) shape. The cells can be short, straight or slender in form. Some are cocco-bacillary. There are three families (Table 6.9) in this section. Two families are catalase positive while the other is catalase negative and separation is on aerobicity, growth temperatures and motility.

Table 6.9

Differentiation of the Families in Section 14

	Mot	Aerb	G35°C	Cat	H_2S	Agluc
A *Lactobacillus*	-	FA	+	-	-	+
A *Erysipelothrix*	-	FA	+	-	+	+
B *Brochothrix*	-	FA	-	+	-	+
B *Listeria*	+	FA	+	+	-	+
C *Kurthia*	+	SA	+	+	-	-
C *Caryophanon*	+	SA		+		

Mot, motile; Aerb, aerobicity; FA, facultative anaerobe; SA, strict aerobe; G35°C, growth at 35°C; Cat, catalase; H_2S, hydrogen sulfide produced; Agluc, acid from glucose.

Members of this section are commonly mesophilic O.J.-III requiring complex media for growth. Most are non-pigmented and non-motile but one family is motile by peritrichous flagella. The families incorporated into Section 14 are: (Family A) catalase negative facultative anaerobes, (Family B) catalase positive facultative anaerobes, and strict aerobes (Family C).

Lactobacillus

This genus belongs to family A and strains are found in a variety of habitats from acidic organic wastes to animal intestines and buccal cavities. Some species are used in the production of yogurts and also play major roles in the generation of flavors during the maturation of cheeses. Cells are often long, slender (occasionally bent) rods, while others are short, often coryneform, or cocco bacilli. It is very usual for the cells to form into chains. While species are all gRAM positive, the staining is often either bipolar, or there may be an internal barring or granulation within the cells. The metabolism is fermentative metabolism with at least half of the carbon end product being lactate. Differentiation of the species of *Lactobacillus* is given in Table 6.10.

Table 6.10

Species Differentiation in *Lactobacillus*

	Ac	Ae	Al	Am	Ar	Ag	Grp
delbrueckii	+/-	-	-	-	-	-	I
lactis	+ /-	+	+	-	-	+/-	I
bulgaricus	-	-	+	-	-	-	I
acidophilus	+	+	+	-	+/-	+	I
amylophilus	-	-	-	-	-	+	I
amylovorus	+	+	-	-	-	+	I
helveticus	-	-	+	-	-	+	I
jensenii	+	+	-	+/-	-	+	I
ruminis	+	+	+/-	-	+	+	I
salivarius	-	+/-	+	+	-	+	I
agilis	+	+	+	+	+	+	II
casei	+	+	+	+	+		II
plantarum	+	+	+	+	+	+	II
sake	+	+	+	+	-	+	II
coryneformis	-	-/+	+/-	+	-	+	II

Ac, acid from cellobiose; Ae, acid from esculin; Al, acid from lactose; Am, acid from mannitol; Ar, acid from raffinose; Ag, acid from galactose; Grp, grouping for *Lactobacillus.*

Most are microaerophilic but surface growth can be enhanced by the elevation of the CO_2 in the atmosphere to 5%. All strains are catalase negative. Pigmentation is rare. When it occurs, the range of colors can be from yellow to orange to rust to brown. All are O.J.-III and complex nutritional requirements are commonly needed to culture strains of this genus. Facultative psychrotrophy is common and many strains are aciduric with optimal pH ranging from 5.0 to 6.2.

Erysipelothrix

This genus also belongs to family A and is commonly parasitic in mammals, birds and fish but some strains are pathogenic to mammals and birds. For example, *E. rhusiopathiae* causes swine erysipelas in pigs. Cells are straight or slightly curved rods with a tendency to form long filaments but may also occur singly or in short chains. In paired arrangements, the cells may become angled to give a "V" form and are mesotrophic. Strains will ferment glucose and a limited range of other carbohydrates to acid but without gas. As fastidious O.J.-III, they require growth factors.

Brochothrix

A family B genus, *Brochothrix* has unbranched regular rods occurring singly, in short chains or long filamentous chains which can form into "knotted" masses. While gRAM positive, some older cells lose the ability to retain the stain and appear to be gRAM negative. Many are facultatively psychrotrophic. Fermentative with L(+) lactic acid being a major product, no gas is produced. These are fastidious O.J.-III with some organic growth factors required.

Listeria

Belonging to family B, members of this genus are widely distributed. Species are found in water, mud, sewage, vegetation and in the feces of animals. *L. monocytogenes* can be pathogenic to humans causing a range of conditions from meningitis, encephalitis, septicemia, endocarditis, and abortion to abscesses. Cells are regular short rods with rounded ends. Some cells may be curved and older cultures can include filamentous forms. Facultatively psychrotrophic but with optimal temperatures at 30 and 37°C. All strains are catalase positive and oxidase negative. Species differentiation is given in Table 6.11. Glucose fermentation products are dominated by L(+) lactic acid. Growth is generally over a narrow pH range of 6 to 9 but strains are salt tolerant and will grow in 10% NaCl supplemented nutrient broth.

Table 6.11
Species Differentiation in *Listeria*

	Camp	Betah	Am	Ax	Hh	Nr	VP
monocytogenes	+	+	-	-	+	-	+
innocua	-	-	-	-	+	-	+
ivanovii	-	-	-	+	+	-	+
grayi	-	-	+	-	-	-	+
murrayi	-	-	+	-	-	+	+
denitrificans	-	-	-	+	-	+	-

Camp, CAMP test using 5% sheep's blood and *Staphylococcus aureus*; Beta, beta heamolysis; Am, acid from mannitol; Ax, acid from xylose; Hh, hydrolysis of sodium hippurate; Nr, nitrate reduction; VP, Voges-Proskauer test.

Kurthia

A genus in family C has been recovered from a diverse variety of sources including foods, soils and fecal material. Strains possess the rod - coccoid life cycle and so that while the young cultures consist of regular unbranched rods, the cultures mature to become dominated by coccoid cells. The rods are motile with peritrichous flagella but some strains are non-motile. All are strictly aerobic and oxidase negative and species differentiation is given in Table 6.12. Carbohydrates are not degraded to acidic products in peptone media and all strains are O.J.-III.

Table 6.12
Species Differentiation of *Kurthia*

	G40°C	Ag	Hh	Am
zopfii	-	-	+	-
gibsonii	+	-	+	-

G40°C, growth at 40°C; Ag, acid from glucose; Hh, hydrolysis of sodium hippurate; Am, acid from mannitol.

Caryophanon

A genus belonging to family C, species have only been isolated from soil freshly contaminated with cattle dung or the dung itself. The cells are straight or slightly curved multicellular rods (i.e., trichomes). The ends are rounded and the trichomes may be attached in short chains. Motile by peritrichous flagella, they are all mesotrophic but they have proven to be difficult to cultivate in the laboratory. Acetate is the only major carbon source. Catalase is strongly positive.

Section 15, Irregular Non-Sporing gRAM Positive Rods

A large and diverse collection of gRAM positive rods which bear the common characteristic of having an irregular shape to the cell. This shape may be a result of swellings along the length of the cell or club-shaped ends. There is a whole range of types from aerobic and facultatively anaerobic to anaerobic genera. In this section, there are three families based upon aerobicity: Family A, strictly aerobic; Family B in which the vast majority are facultative anaerobic; and Family C containing the strict anaerobes. Differentiation of many of the genera in the three families is given in Table 6.13. Selected genera will be described for each family.

Table 6.13
Differentiation of Genera in Section 15, gRAM Positive Irregular Rods

	Fam	Cat	R-C	Gluc	Mot
Corynebacterium	A/B	+	-	R/-	-
Gardnerella	B	-	-	A	-
Arthrobacter	A	+	+	-	-
Curtobacterium	A	+	-	R	+
Caseobacter	A	+	+	R	-
Microbacterium	A	+	-	R/A	-/+
Cellulomonas	B	+	-	A	+/-
Arachnia	B	-	-	A	-
Propionibacterium	B/C	+	-	A	-
Eubacterium	C	-	-	A	-
Acetobacterium	C	-	-	-	+
Actinomyces	C	+/-	-	A	-

Fam, family based upon strict aerobes (A), facultative anaerobes (B) and strict anaerobes (C); Cat, catalase; R-C, rod to coccus life cycle; Gluc, glucose respired (R) or fermented to acid (A); Mot, motility.

Arthrobacter

This genus in family A is very commonly found in soils and has been found to form the largest non-mycelial component in many soils. These bacteria possess a very strong rod - coccus life cycle, are gRAM positive and strictly aerobic with some strains occasionally motile. During the life cycle, the coccoid elements form during the stationary growth phase. When subcultured, these coccoid cells produce one or more protruberances that grow out from the cell and form into the typical rod shaped cells. Branching may occur and some cells may become arranged in the "V" formation. Most strains are O.J.-II normally with very simple nutritional needs but some strains do require biotin for growth (O.J.-III). Generally, strains grow reasonably well on yeast-peptone agar to form 3 to 5 mm diameter colonies which are usually cream to buff in color but occasionally yellow.

Microbacterium

In family A, species are often isolated from milk and dairy products (*M. lacticum*) while others are isolated from raw sewage and activated sludge (*M. laevaniformis*). In this genus, the young cells are long, slender and irregular rods that often become arranged in the "V" formation. As the culture matures, the rods become shorter and the proportions become more coccoid-like. The frequency of this occurrence suggests that this is not a true rod - coccus cycle. While this genus is strictly aerobic, some strains are able to grow weakly under anaerobic conditions. Most strains are non-motile but the few strains that are motile possess one to three flagella. All strains are catalase positive and fastidious O.J.-III requiring the B-vitamins and some amino acids as well. Species are differentiated in Table 6.14. Most grow well on yeast extract-peptone-glucose agar and typical colonies are circular, opaque and glistening often with yellowish pigments.

Table 6.14

Species Differentiation in *Microbacterium*

	Pig	Mot	Nit	H_2S
lacticum	y - w	-	+	-
imperiale	r - o	+	-	-
laevaniformis	y	-	-	+

Pig, pigment; y, yellow; w, white; o, orange; Mot, motility; Nit, nitrate reduced; H_2S, hydrogen sulfide produced from cysteine.

Corynebacterium

Perhaps this is the major genus in families A and B and includes a broad range of animal and plant parasites some of which are pathogenic. Two separate listings for the animal and plant parasites are presented in the Bergeyian system. Of the animal parasites, some of the pathogenic species include: animals, *C. diphtheriae* (produces a lethal exotoxin and causes diphtheria in humans) and *C. renale* (cystitis and pyelitis in cattle). In plants, some of the pathogens are *C. michiganense* (vascular wilt, canker, leaf spot and fruit spot in plants belonging to the *Solanaceae*) and *C. sepedonicum* (vascular wilt and tuber rot in the potato). As well as the parasitic species, there is also a large and diverse range of strains. While some are strictly aerobic, others are facultatively anaerobic. Animal pathogenic species are differentiated in Table 6.15 while the plant pathogens are differentiated in Table 6.16. Cells are straight to slightly curved rods with tapered ends. Club shaped forms may be observed often but ellipsoid, ovoid and more rarely "whip handle" forms may be seen. Cell division is of a unique "snapping" type in which the cells come to lie in angular and palisade arrangements. The gRAM positive reaction often involves the stain being taken up very unevenly in the cell.

Brevibacterium

Belonging in family A, this species has been isolated from the skin, the surfaces of ripening cheese, marine fish, and various pig and poultry manures. Many strains are salt tolerant, non-motile, short, and have unbranching rods. The life cycle is, however, often complex. For example, *B. linens* is coccoid but has outgrowths of rod shaped cells that elongate with "V" forms before the cycle returns to the coccoid form. Pigmentation is often common with a form of light-induced carotenoid pigments and only appears after prolonged exposure to light.

Table 6.15

Species Differentiation of Animal Pathogens, *Corynebacterium*

	Ag	Agl	Am	Amt	As	Hh	Hc	Nr
diphtheriae	+	+	+	+	+	-	-	+
pseudotuberculosis	+	+	+	+	-	-	-	+/-
xerosis	+	+	+	-	+	+	-	+
pseudodiphtheriticum	-	-	-	-	-	+	-	+
kutcheri	+	-	+	+	+	+	-	+
renale	+	-	+	+/-	-	+	+	-
cystidis	+	-	-	+	-	+	-	-
pilosum	+	-	+	+	-	+	-	+
flavescens	+	+	+	-	-	-	-	-
bovis	+	+	-	+	-	+	-	+

Ag, acid from glucose; Agl, acid from galactose; Am, acid from mannose; Amt, acid from maltose; As, acid from salicin; Hh, hydrolysis of sodium hippurate; Hc, hydrolysis of casein; Nr, nitrate reduction.

Table 6.16
Species Differentiation of Plant Pathogens, *Corynebacterium*

	Amn	Amt	Inu	Uc	Asc	H_2S	Lv	Gel
michiganense	+	-	-	+	+	+	-	+/-
insidiosum	+	-	-	-	-	-	-	-
iranicum	+	-	-	-	+	+	-	-
nebraskense	+	-	-	+	+	+/-	+	-
sepedonicum	+/-	+	-	+	+	-	-	-
tritici	+	+	+	+	+	+	+/-	-
rathayi	-	+	-	-	+	+	+	+

Amn, acid from mannose; Amt, acid from mannitol; Inu, utilizes inulin; Uc, utilizes acetate; Asc, acid from salicin; H_2S, hydrogen sulfide from peptones; Lv, levan produced; Gel, gelatin proteolysis.

Cellulomonas

This genus, a member of family B, is isolated from soils frequently and occasionally from decaying vegetative material. This genus plays a major role in the degradation of cellulose. Young cultures are dominated by slender irregular rods which may be angled to each other and appear as "V" forms. Older cultures are dominated by shorter rods but some coccoid cells are also present. While gRAM positive, cells decolorize very easily to appear negative. Motile by one polar or subpolar, or several lateral flagella. Facultatively anaerobic, glucose is fermented to acid under both aerobic and anaerobic conditions; catalase is positive. The unique feature of this genus is that all strains are cellulolytic. They are also capable of reducing nitrate to nitrite, and frequently have weak protease and amylase systems. Species differentiation is given in Table 6.17. Cultures grow reasonably well on peptone-yeast extract agar giving opaque convex yellow colonies.

Table 6.17
Species Differentiation of *Cellulomonas*

	Mot	Urb	Urf	Ult	Uac	Ula
flavigena	-	+	-	-	+	-
biazotea	+	-	+	+	+	-
fimi	+	-	-	+	-	+
gelida	+	-	-	-	+	+/-
uda	-	-	-	-	+	+
cellasea	-	-	-	+	+	-

Mot, motility; Urb, utilization of ribose; Urf, utilization of raffinose; Ult, utilization of lactate; Uac, utilization of acetate; Ula, lactose utilization.

Propionibacterium

This genus belongs to family B and species are isolated from milk and dairy products and also from a number of sites in the human body including the skin and subcutaneous tissues. Cells in this genus are very pleomorphic rods often forming diphtheroid and club-shapes. Cells may have one end rounded while the other end is tapered. Division may cause the cells to split into "bifid" forms and even branch while coccoid cells may also be observed. There may be such a variety of the "V" and "Y" forms that clumps of cells may appear to be in the "Chinese character" formation (cuneiform). Strains are non-motile and gRAM positive; these O.J.-III bacteria are distinguished by the copious amounts of propionic and acetic acids generated during fermentation. While catalase positive, these strains are facultatively anaerobic or strictly anaerobic but aerotolerant. Colonies are white, grey, yellow, pink, orange or red in color.

Actinomyces

The *Actinomyces* is a major genus in family B. Strains are often isolated from various sites on animal and human bodies. Species of particular importance include *A. bovis* (actinomycosis in cattle), *A. israelii* (human cervicofacial, thoracic, and abdominal actinomycosis, and dental plaque), and *A . odontolyticus* (dental plaque and calculus). Cells have a considerable diversity in form ranging from straight or slightly curved rods to slender filaments with true branching. Irregular forms are common with the diphtheroid arrangements of "V," "Y" and "T" forms with palisading occurring. Coccobacillary forms may also be present. Where filaments are produced these may be straight or wavy with or without clubbed, swollen or clavate ends. Although gRAM positive, they often stain very irregularly with beaded or barred appearances occurring frequently. They are facultatively anaerobic with many strains growing better anaerobically. Species differentiation is described in Table 6.18. Mature colonies are often smooth, soft to mucoid or rough and dry to crumbly in texture but aerial growths (filaments) do not occur. Pigmentation is commonly from a white to a grey or creamy white color but a few strains form red pigments. Fermentative O.J.-III, bacteria grow better in enhanced carbon dioxide atmospheres. Acid but not gas is commonly produced from carbohydrates. End products of glucose fermentation include formic, acetic, lactic and succinic acids but not propionic. Strains are variable in the catalase reaction and nitrate reduction.

Table 6.18
Species Differentiation in *Actinomyces*

	Cb	Fl	Cat	Nr	St	Dns	Afc
bovis	+/-	+/-	-	-	-	+	-
israelii	-	+	-	+/-	+/-	-	R,X,T,C
naeslundii	-	+/-	-	+	+/-	-	R,X,T,C
odontolyticus	+/-	+/-	-	+	+/-	+/-	X,T
viscosus	+/-	+	+/-	+	-	+	X
pyogenes	+	-	-	-	+/-	+	X,T,C
meyeri	+/-	-	-	+	-	-	X

Cb, coccobacillary; Fl, filaments; Cat, catalase; Nr, nitrate reduction; St, starch hydrolysed; Dns, DNAase; Afc, Acid from the carbohydrates; R, raffinose; X, xylose; T, trehalose; C, cellobiose.

Eubacterium

Belongs to family C and strains are found in the various cavities of humans and other animals and in plant products. It has been noted to cause infections in soft tissues. Some strains have been found in soils. Obligate anaerobes of variable cell form which produce a mixture of organic acids from carbohydrates or peptones often including large amounts of butyric, acetic and formic acids.

Acetobacterium

This genus belongs to family C and has been isolated from marine- and freshwater sediments and sewage sludge. The cells are oval shaped short rods which are motile by one or two subterminal flagella. They are all strictly anaerobic and can oxidize H_2 and reduce CO_2 (O.J.-I) to form acetic acid. They can also be O.J.-II and ferment some substrates such as fructose but, in all cases, the end products include acetic acid as the main carbon product.

Section 16, *Mycobacteriaceae*

This section is formed by a unique characteristic universally displayed by the strains which feature acid-fastness. The reason for this property of acid-fastness is the waxy materials found in the cell walls of the mycobacteria. This allows the cell to repel acidic materials. While these bacteria have been a focus of much research due to their implication in cases of tuberculosis (historically referred to as consumption), the mycobacteria are widely distributed within the environment and many are not pathogenic. The mycobacteria are comprised of the aerobic, non-motile rod shaped bacteria that are often slow growing and are characteristically strongly acid-fast positive for at least some part of the life cycle. Only a few other genera may be partially or weakly acid-fast. These include *Nocardia* and *Rhodococcus*. These other two genera are normally gRAM positive to a strong degree whereas *Mycobacterium* species are weakly gRAM positive. There is only one family (*Mycobacteriaceae*) containing a single genus.

Mycobacterium

This genus has been well documented because *M. tuberculosis* produces tuberculosis (consumption) in humans and some other mammals. Some of the other species can produce tubercles in a variety of other warm blooded animals, but there are a vast range of saprophytic strains that occur in soils and these are non-pathogenic. Cells are slightly curved or straight rods which may sometimes be seen as either branching, filamentous, or form a transient mycelial-like growth. Where this latter form of growth is disturbed, the mass may fragment into both rod and coccal elements. Species are differentiated in two manners depending upon whether they are slow growing (Table 6.19) or fast growing (Table 6.20). Occasionally, rods may elongate and intertwine around each other.

Table 6.19

Species Differentiation of *Mycobacterium*, Slow Growing

	Ur	Nr	G25°C	G45°C	Pcr	Scr	5%Nc
tuberculosis	+	+	-	-	-	-	-
bovis	+	-	-	-	-	-	-
kansasii	+	+	+	-	+	-	-
marinum	+	-	+	-	+	-	+/-
gastri	+	-	+	-	-	-	-
gordonae	-	-	+	-	-	+	-
simiae	+	-	+	-	+	-	-
scrofulaceum	+	+/-	+	-	-	+	-
avium	-	-	+/-	-	-	-	-
xenopi	-	-	-	+	-	+/-	-

Ur, urease; Nr, nitrate reduction; G25°C, growth at 25°C; G45°C, growth at 45°C; Pcr, photochromogenic, generates pigment in the presence of light; Scr, schotochromogenic, generates pigment only after exposure to light; 5%Nc, grows in 5% NaCl.

The acid-alcohol-fast characteristic is present during at least some stage in the life cycle and so negative results can be common. Repeated staining is sometimes required to confirm that the culture is acid-fast. Some strains are difficult to stain using the standard gRAM stain methodologies but it is generally considered that all strains are gRAM positive. Differentiating strains into fast growing or slow growing is based upon the rate at which growth occurs in laboratory cultures with less than three days to colonial forms being fast, and longer than two weeks as being slow.

Table 6.20
Species Differentiation in *Mycobacterium*, Fast Growing

	G45°C	G52°C	Gmc	Feu	5%Nc	Pig
chelonae	-	-	+	-	-	-
abscessus	-	-	+	-	+	-
fortuitum	-	-	+	+	+	-
chitae	-	-	-	+/-	+/-	-
smegmatis	+	-	-	+	+	-
phlei	+	+	-	+	+	+
flavescens	+/-	-	-	-	+	+
sphagni	-	-	-	-	-	+

G45°C, growth at 45°C; G52°C, growth at 52°C; Gmc, growth on MacConkey agar; Feu, iron uptake; 5%Nc, grows in 5% sodium chloride; Pig, pigmented.

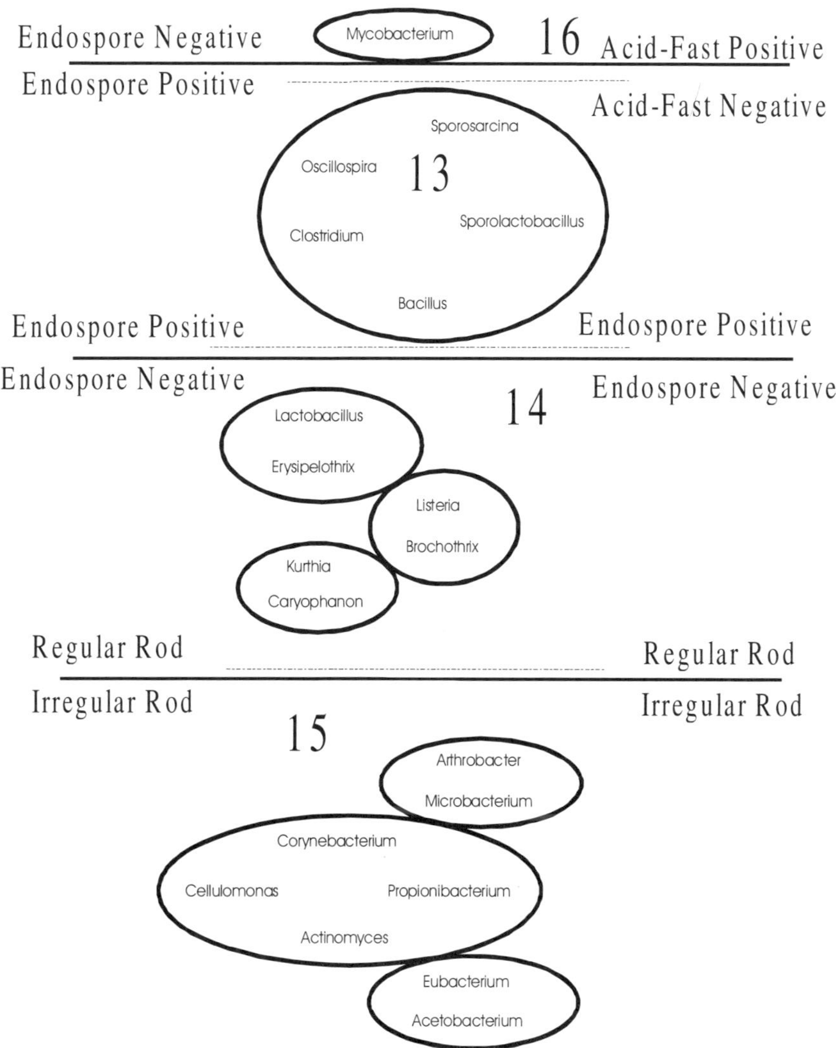

Figure 6.1 Global differentiation of the gRAM positive rods embodied in Sections 13, 14, 15 and 16. The boundaries of the four sections of this figure are defined by the typifying characteristic of each section with continuous lines used to separate each section.

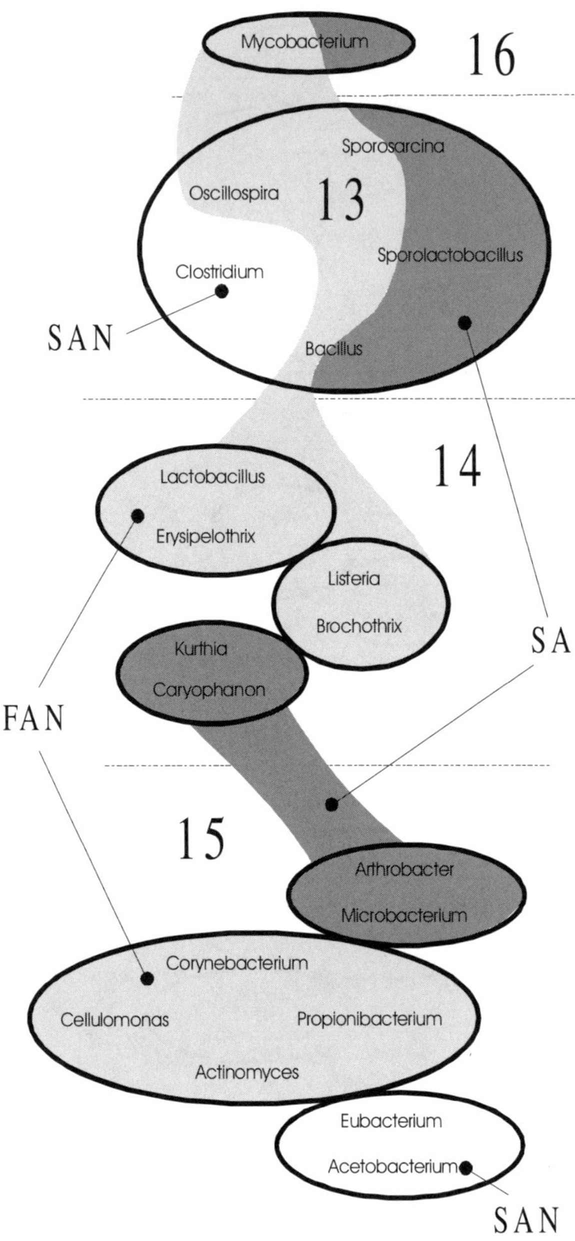

Figure 6.2 Shade plotted diagram illustrating the aerobicity of the gRAM positive rods defined by Sections 13, 14, 15 and 16 based upon their aerobicity. The strictly aerobic (SA, dense shade) are separated from the facultatively anaerobic bacteria (FAN, light shade) and the strictly anaerobic bacteria (SAN, not shaded).

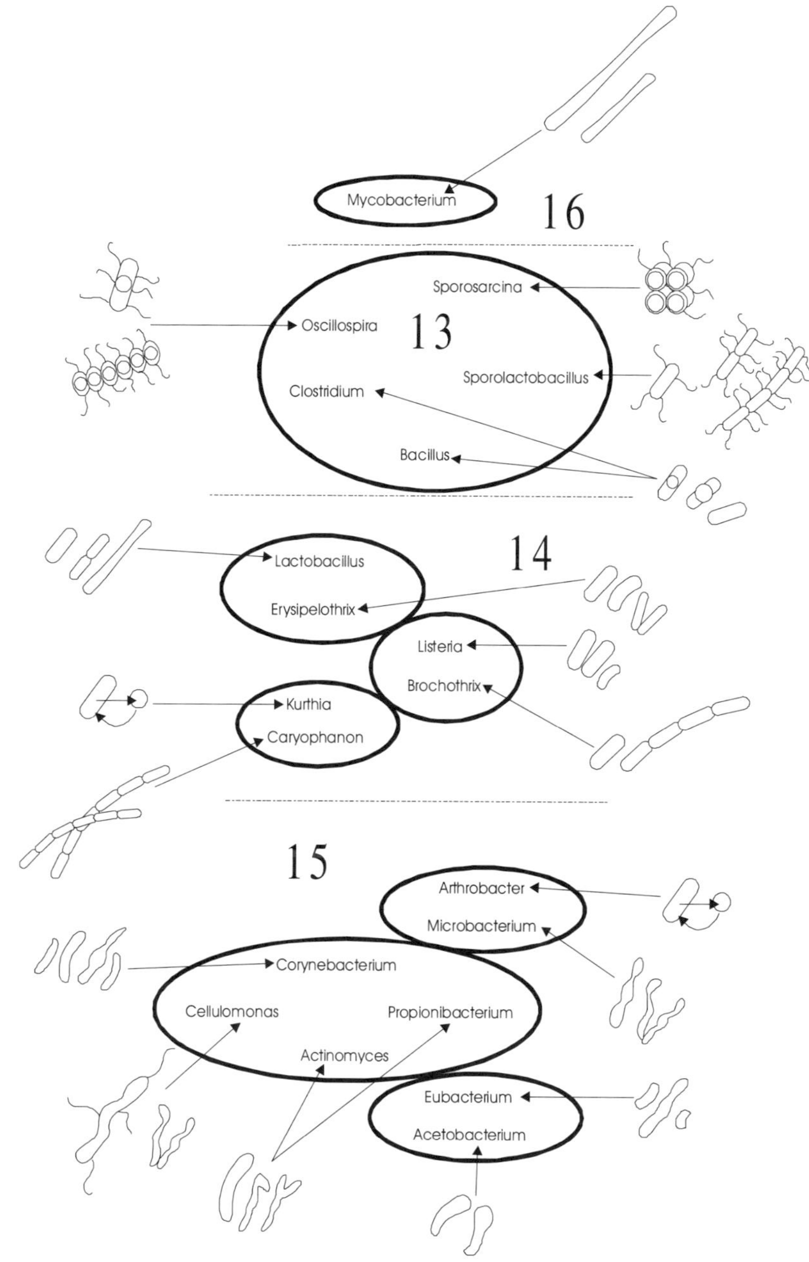

Figure 6.3 Illustrations of the typical cell forms for the various genera embraced within Sections 13, 14, 15 and 16. The forms are defined using lines to indicate which genera possess the feature.

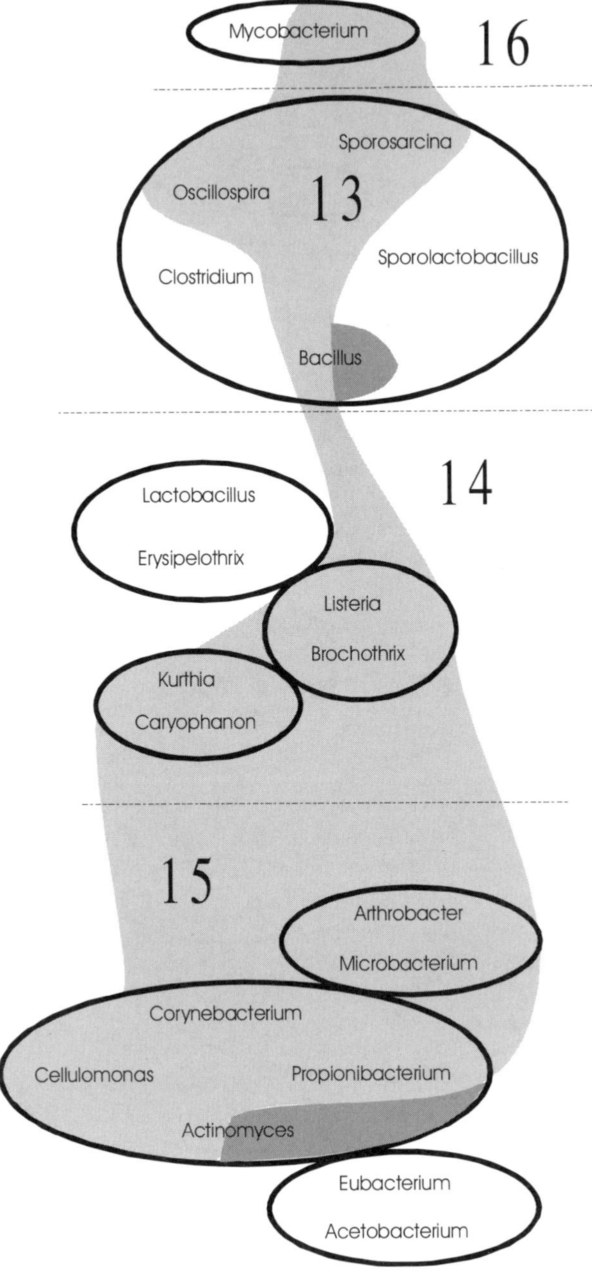

Figure 6.4 Occurrence of the catalase enzyme systems in the gRAM positive bacteria found in Sections 13, 14, 15 and 16. The catalase positive genera are light shaded. Those species within genera that possess the catalase positive characteristic which is lost upon maturation are indicated by the dense shade.

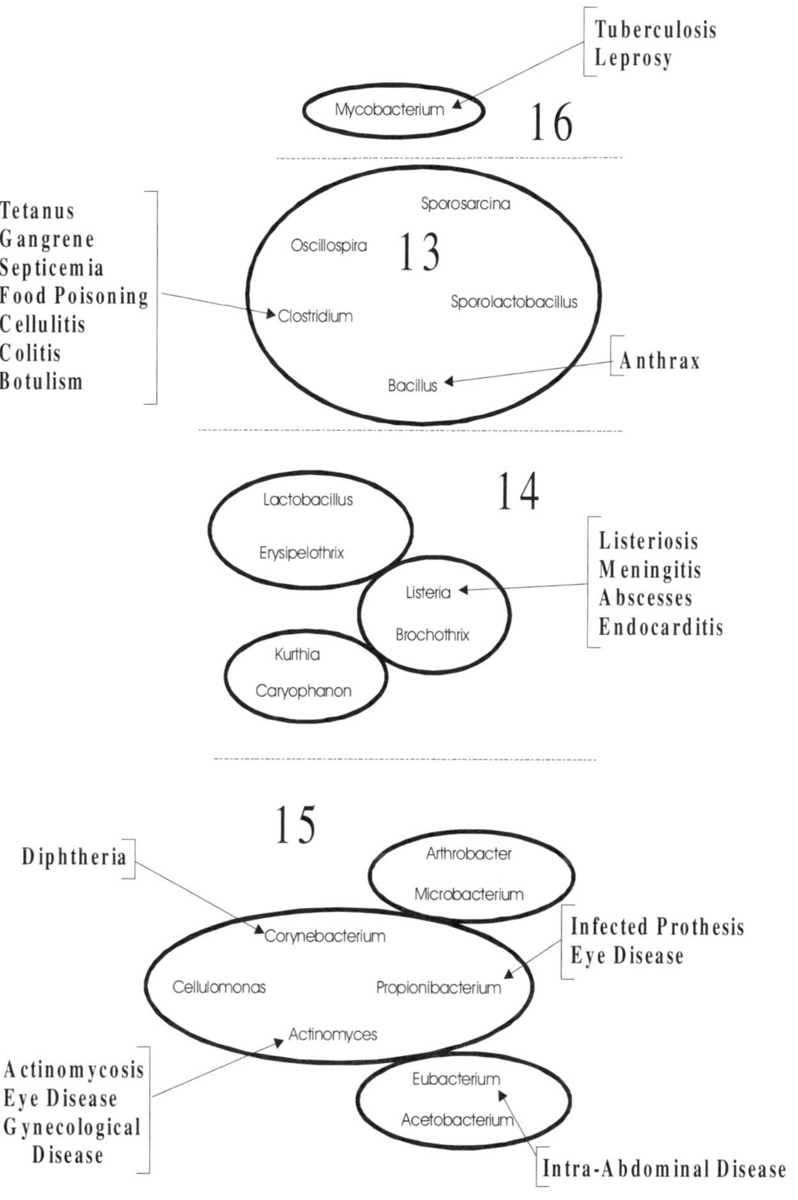

Figure 6.5 Diseases in humans caused by members of Sections 13, 14, 15 and 16, the gRAM positive rods. A separate box is displayed for each genus known to be pathogenic to humans. An arrow with a continuous line connects the box to the name of the genus responsible for those diseases.

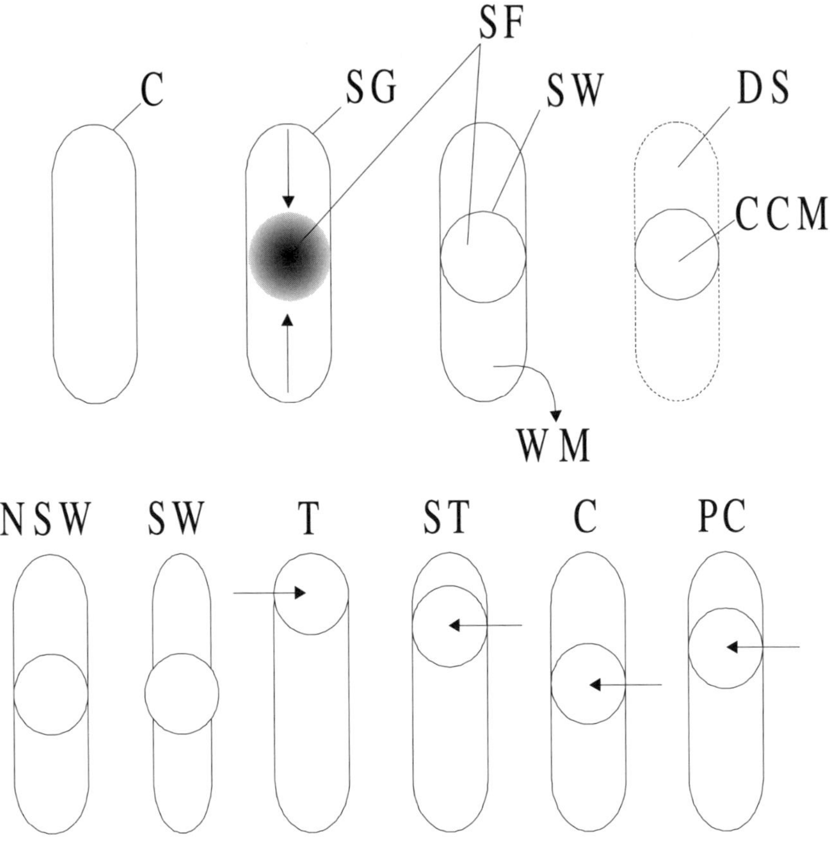

Figure 6.6 Diagrammatic presentation of the formation of an endospore within a cell (C) that becomes a sporangium (SG) within which the spore forms (SF) while the sporangium ejects waste materials (WM). A spore wall (SW) separates the condensed cell mass (CCM) from the disintegrating sporangium (DS) that will now stain gRAM negative. Below is arranged the various defined sizes and positions seen for endospores: NSW, not swelling; SW, swelling the sporangium; T, terminal; ST, subterminal; C, central; and PC, paracentral.

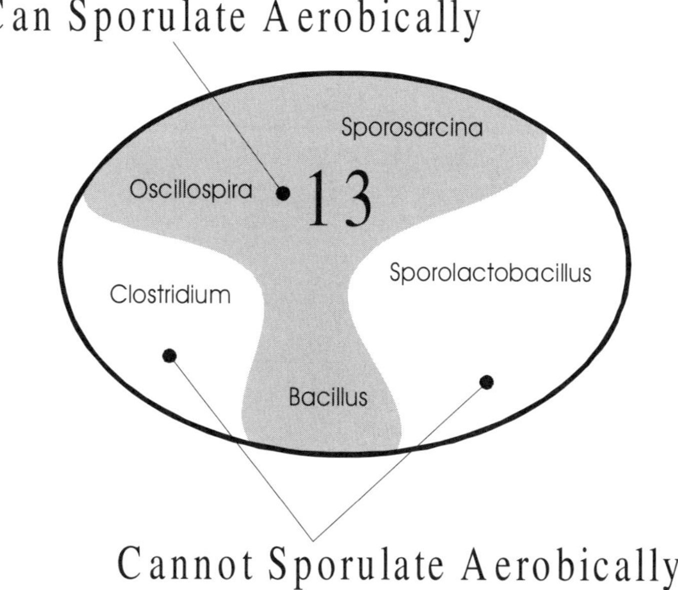

Figure 6.7 Differentiation of the Section 13 endosporogenous bacteria by the conditions under which these spores are formed. The shaded genera are those that are capable of sporulating aerobically while the unshaded genera cannot sporulate under aerobic conditions.

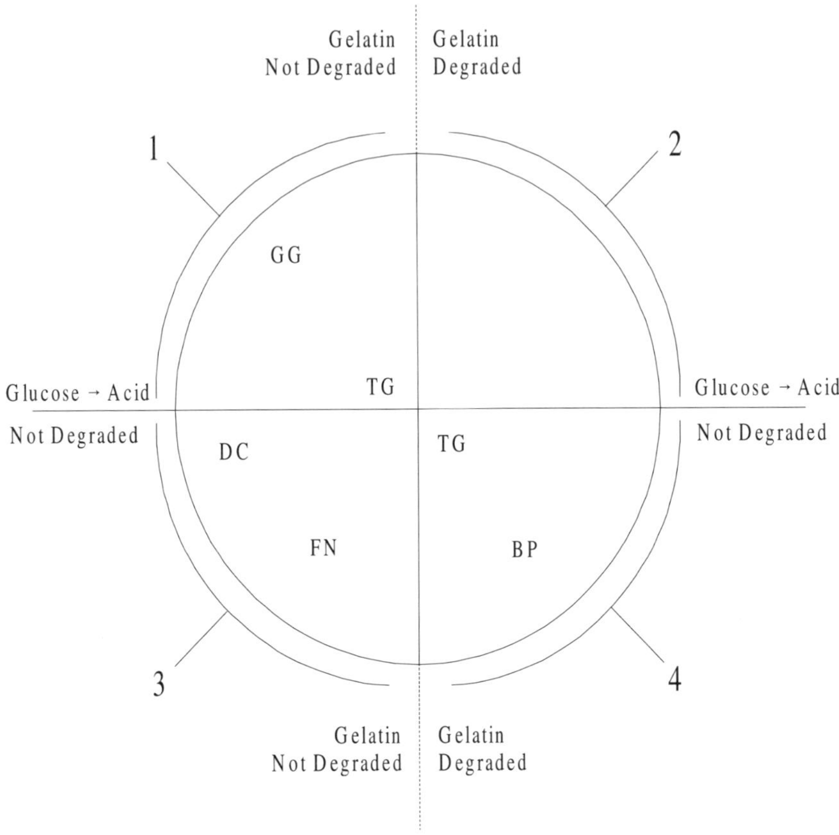

Figure 6.8 Diagrammatic representation of the four major species groups of *Clostridium* based upon their ability to generate gas gangrene (GG), are toxigenic (TG), fixate nitrogen (FN), degrade cellulose (DC) and are industrially exploited to produce butanol (BP).

7

Mycelial Bacteria (Sections 26, 27, and 29)

Bordering the fungi in complexity, these sections of bacteria represent some of the most successful bacteria colonizing soils. Their advantage over other bacteria is that they form distinct mycelial masses (Figure 7.1) very similar to those seen in the fungi. The mycelium is a web-like mass of filaments which form a very stable structure that allows the bacteria to dominate a large surface area. In soils, this filamentous mass can cross, bind and integrate soil particles and so improve the soils' texture. In these bacterial sections, the mycelium is minimally stable for at least a part of the life cycle. Another divergence from the other bacteria is that the mycelial bacteria will produce, on occasions, exospores and specialized filaments to support these sporulations. These exospores as the prefix suggests are formed on the outside (i.e., exo-) of the cells. The role of these exospores is dual. First, the spore forms a method of reproduction since they are often produced in large numbers from a few specialized cells. Second, the exospores are more resistant to adverse conditions than the vegetative cells which means that the exospores form survival mechanisms against extreme environmental conditions. The net effect of this is that very often samples may contain very large numbers of these exospores that are essentially in a "survival" state and do not reflect accurately the true amount of vegetative biomass in the sample (i.e., a false positive or over-estimation is achieved). Sometimes the exospores can also be motile by flagella.

The classification of the mycelial bacteria is complex and so only four of the more significant genera from these sections will be described (Table 7.1, Figure 7.2). Each have some unique characteristics and play a major role in the environment.

Frankia

Species of this genus have become recognized as significant mycelial bacteria somewhat paralleling the role of *Rhizobium* in that it causes both nodulation and nitrogen fixation in the host plant being infested. However, the host plants for *Frankia* are a range of woody plants and the role is more balanced between the parasite and the host. In *Rhizobium*, the plant dominates and incorporates the bacterial cell components into the nodule. With *Frankia*, this dominance does not occur.

The result is a symbiotic filamentous mycelium that induces, and then lives within the root nodules in a wide variety of non-leguminous (commonly woody) dicotyledonous plants. Once formed, these nodules are capable of fixating molecular nitrogen. As well as living in a symbiotic relationship with these plants, the bacteria can also exist in the free state within the soil. A true cellular (septate) mycelium is produced with the branching occurring anywhere along the hyphal filaments. When the nodule is actively fixing nitrogen, the center of the host cell becomes filled with a mass of the hyphae. Pleomorphism is more common at the time that nitrogen fixation is taking place. In addition, globular vesicles are commonly found close to the plant cell walls during the fixation phases.

Table 7.1
Differentiation of Four Selected Genera of the Mycelial Bacteria

	Nfx	Pp	Ap	Aop	Ah	Ms
Frankia	+	+	-	-	+	+
Streptomyces	-	-	-	-	-	+
Dermatophilus	-	-	+	-	-	+/-
Nocardia	-	-	-	-/+	+	-

Nfx, nitrogen fixation; Pp, plant parasite, Ap, animal pathogen; Aop, animal opportunistic pathogen; Ah, aerial hyphae; Ms, mycelium stable.

Streptomyces

Strains are very common in soils with many of the *Streptomyces* producing geosmin-like volatile substances. It is these substances that give the earth its characteristic earthy-musty odor. They form a major part of the microbial biomass of soils and retain that position in part by the copious production of a diverse range of antibiotics. Some of these antibiotics have been exploited such as amphotericin B, chloramphenicol, erythromycin, neomycin, nystatin, streptomycin, and tetracyclin. This genus generates an aerial hypha (filament extending upwards into the air over the mycelial mat). These hyphae divide to form exospores often in chains of 5 to 50 or more non-motile exospores. The surface textures of these spores range from smooth to spiny and warty. All of the strains are strictly aerobic and species are differentiated on the basis of the characterization of the mycelial mat and the form of the exospores.

Dermatophilus

Some species have been extensively studied because some members of this genus are pathogenic to humans causing skin lesions (e.g., scabies). The gRAM positive cells grow as filaments that may then form into either a mycelium or a thallus that have both transverse and longitudinal cell wall division. Sometimes this can give the appearance of a brick wall (muriform) or just simply an irregular mass of cells. Cells may be released from these masses and characteristically become motile. Generally, high concentrations of CO_2 are required to generate growth and formation of the cellular masses.

Nocardia

This is a genus that is widely distributed in soils but there are some strains that are opportunistic pathogens to animals and humans. Species differentiation is in Table 7.2.

Table 7.2

Species Differentiation in *Nocardia*

	Ag	As	Ai	Aa	Am	Agl	Dc	Du
asteroides	+	-	-	-	-	+/-	-	+
farcinica	+	-	-	-	-	+/-	-	+
brasiliensis	+	-	+	-	-	+	+	+
otitidis-caviarium	+	-	+	-	-	-	-	+
amarae	+	-	+	-	+	-	-	+
brevicatena	-	-	-	-	-	-	-	-
carnea	+	+	+/-	-	-	+	-	-
vaccinii	+	+/-	+/-	-	+/-	+	-	+
transvalensis	+	+/-	+/-	+	-	+	-	+

Ag, acid from glucose; As, acid from sorbitol; Ai, acid from inositol; Aa, acid from adonitol; Am, acid from maltose; Agl, acid from galactose; Dc, decomposition of casein; Du, decomposition of urea.

Nocardia tend to growth both on, and penetrating under, the surface of the supporting media with the formation of rudimentary to extensively branching vegetative hyphae. These growths are commonly fragile and disrupt easily to yield bacteroidal, rod shaped and coccoid cells. Aerial hyphae are commonly formed but may only be viewed microscopically. Conidia are formed but no sporangial, sclerotial or other specialized spore structures are formed. Cells are gRAM positive to gRAM variable and there are some strains that are weakly acid fast at some stages in the life cycle and are catalase positive.

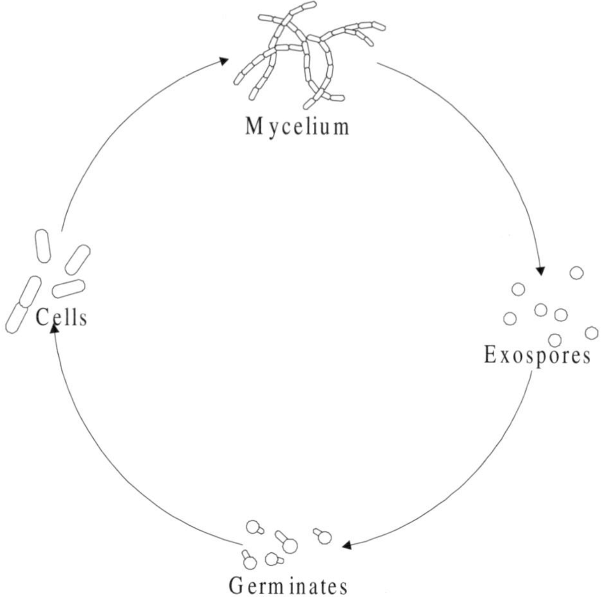

Figure 7.1 Typical life cycle observed in the mycelial bacteria belonging to Sections 27, 28, and 29.

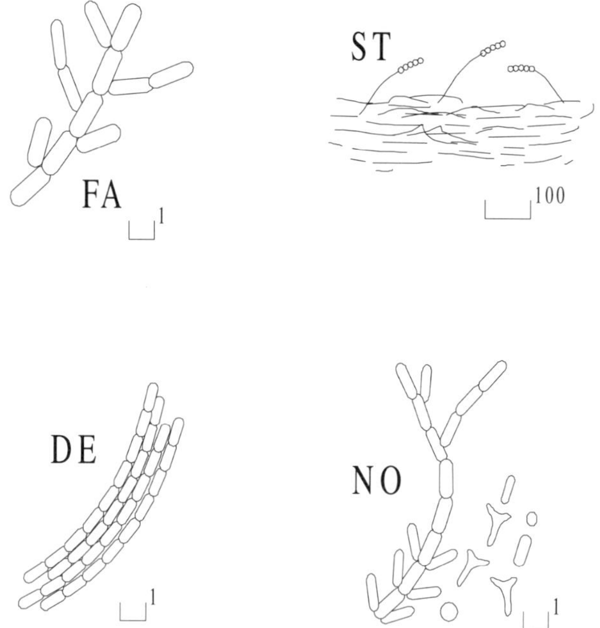

Figure 7.2 Illustration of the cell forms for the mycelial bacteria belonging to the genera: *Frankia* (FA), *Streptomyces* (ST), *Dermatophilus* (DE), and *Nocardia* (NO). Scales are in microns.

8

Other Bacterial Sections

This chapter covers a range of other bacterial sections that are important under various environmental and pathological conditions. Three main groups of gRAM negative bacteria are included in this chapter including the spiral and helical forms (Sections 1, 2 and 3), the prosthecate bacteria that bear extensions out of the cells (Section 21) and the gliding bacteria that are capable of propulsive forms of movement without the use of flagella (Sections 23 and 24). Each of these three groups of bacteria are clearly distinguishable and the features are defined at the start of the description for each of these groups of bacteria.

Sections 1, 2, and 3, Helical and Spiral gRAM Negative Bacteria

The terms "helical" and "spiral" mean that the cells are bent to form a partial or complete single spiral (Figure 8.1). Therefore, all possess the common characteristic that the cell is not straight but helical bearing one or more complete harmonic cycles along the axis of the cell. The cells are mostly unicellular forms that are either gRAM negative or gRAM zero. They are separated (Figures 8.2 and 8.3) into three families: helically shaped, motile bacteria bearing an outer sheath (Section 1), microaerophilic, motile, gRAM negative vibrioid to comma shaped bacteria (Section 2) and nonmotile, gRAM negative, curved bacteria (Section 3). The three sections are differentiated in Table 8.1 (see also Figures 8.2 and 8.3). A number of genera are pathogenic (Figure 8.4).

Table 8.1
Differentiation of the Sections for Helical and Spiral Bacteria

	Cell morph	Aerobicity	Motility
Section One	Hel, Spir C	SN,FA,SA	+
Section Two	Vibr to C	SA, MA	+
Section Three	Comma	SA	-

Cell morph, cell morphology; Hel, helical; Spir, spiral; C, curved; Vibr, vibrioid; Comma, comma shaped; SN, strict anaerobic; FA, facultative anaerobe; SA, strict aerobe; MA, microaerophilic.

Section 1, Motile Spiral or Helically Curved Bacteria

There are four major genera in this section that are differentiated in Table 8.2. This is based upon aerobicity, whether the strains are free living or parasitic and cell size.

Table 8.2
Differentiation of the Genera in Section 1

	FL	SA	FA	SAN	HA	CL
Spirochaeta	+	-	+	+	-	5 - 250
Treponema	-	-	-	+	+	5 - 20
Borrelia	-	+	-	-	+	3 - 20
Leptospira	+	+	-	-	+	12

FL, free living; SA, strict aerobe; FA, facultative anaerobe; SAN, strictly anaerobic; HA, host associated; CL, cell length (microns).

Spirochaeta

Strains are indigenous in aquatic systems and associated sediments and found in both marine and freshwater environments. None are known to be pathogenic. The cells are helical cells and less than 1 micron in diameter but up to 250 microns in length. Most strains have two periplasmic flagella (flagella which wind around the body of the cell before extending beyond the cell). Under stress, these cells form spherical structures 0.5 to 2.0 microns in diameter. Cells are able to perform locomotory functions when suspended in liquids and can also "crawl" or "creep" when in contact with solid surfaces. Strains are either strictly anaerobic or facultatively anaerobic. Under aerobic conditions, most strains are able to generate a range of carotenoid pigments giving yellow, orange or red colors to the colonies. They are able to use a variety of carbohydrates as both carbon and energy sources with the main products of carbohydrate metabolism anaerobically being ethanol, acetate, hydrogen and carbon dioxide.

Treponema

Members of this genus are found in the oral cavity, intestinal tract and genital areas of many animals including humans. *T. pallidum* is of particular concern because it causes syphilis in humans. This species has a very distinctive cell that often appears to resemble a tightly wound coiled spring. Although of small diameter, these helical rods can range from 5 to 20 microns in length. They bear one or more rather unique periplasmic flagella. These are flagella that have an axial filament or axial fibril inserted at each end of the cell. The cells can also be complex with fibrils often observed as tubules within the cell. This usually happens closer to the cell walls. When the cells are subjected to stresses, it is common for the cells to become more spherical in shape. This is particularly common in old cultures. Because of the complex structures in and around the cells, most strains stain poorly using the standard gRAM methodology and better results (e.g., observing cell form) are obtained using either dark-field or phase contrast microscopy. In liquid media, cell movement is commonly rotational but on semi-solid or solid media, the movement becomes "serpentine." All are O.J.-III and are capable of using carbohydrates or amino acids as the primary sources of energy and carbon. These bacteria are strictly anaerobic although some have been found to grow under microaerophilic conditions. Cultivation has proved to be difficult for some strains particularly in artificial media.

Leptospira

This genus is found in diverse habitats including soil and waters. Some strains are parasitic on animals including humans. For example, *L. interrogans* can produce subclinical to lethal infections in a wide variety of animals. Frequently, these infections will localize in the kidneys before spreading to other organs and tissues. Other strains are recovered from waters and, due to their flexible nature and small diameter, they can pass through some of the standard membrane filters. This is because the cells are very flexible spirally shaped rods that have a very small diameter (0.1 microns) and lengths of 6 to greater than 12 microns. All strains are strictly aerobic and these gRAM negative cells are often difficult to observe except by dark-field or phase contrast microscopy because of their small size. All strains are both catalase and oxidase positive. Most are O.J.-III and utilize fatty acids or fatty alcohols as the sources of energy and carbon.

Section 2, Motile Aerobic Helical and Vibrioid gRAM Negative Bacteria

There are a range of helical and spiral gRAM negative bacteria that are motile and aerobic. These are differentiated in Table 8.3 and three of the genera are described in detail below.

Spirillum

This genus is found in many stagnant freshwater environments including distilled and deionized water containers. The cells are rigid helical shapes that have a relatively large diameter (1.4 to 1.7 microns) and length of 14 to 60 microns. The cells have complex structures including a polar membrane underneath the cell membrane and there are bipolar tufts of flagella that allow the cells to be motile. Strains are usually microaerophilic and thrive in waters with lower oxygen concentrations. While the strains are aerobic, they are catalase negative although oxidase positive. They are very difficult to culture on solid media except under very carefully controlled conditions and they are inhibited by even trace amounts of peroxides and will not tolerate NaCl at concentrations greater than 0.02%. Phosphate concentrations exceeding 0.01 M are also inhibitory. Carbohydrates cannot be used as a source of carbon but salts of some organic acids such as succinate can be used. Vitamins are not required for growth.

Table 8.3

Differentiation of Genera in Section 2

Motile Helical and Vibrioid Aerobic and Microaerophilic Bacteria

	Cat	Oxi	Pbh	CO_2r	Cell shape
Aquaspirillum	+	+	+	-	helical
Spirillum	-	+	+	-	helical
Azospirillum	+	+	+	-	vib/c
Campylobacter	+/-	+	-	+	vib/c
Bdellovibrio	+	+	-	-	vib/c
Spirosoma	-	+	-	-	vib/c/s

Cat, catalase; Oxi, oxidase; Pbh, poly B-hydroxybutyrate present; CO_2r, carbon dioxide required for growth; vib, vibrioid; c, curved; s, straight.

Bdellovibrio

Species of *Bdellovibrio* occur widely in soils, waters, and sewage as predators on other bacteria. They mainly target other gRAM negative bacteria such as *Escherichia* and *Spirillum*. When a suitable target cell falls into range, the *Bdellovibrio* cell literally rams into the target with sufficient force to tear the cell wall and propel the invading cell into the victim cell! Once inside the cell, growth begins (Figure 8.5). The cells of *Bdellovibrio* are, not surprisingly, small and comma-shaped with a single polar flagellum. This flagellum is unusual in that it is surrounded by a sheath which is continuous with the outer walls of the cell. There are two phases in the life cycle. In phase one, the cell is predatory on other bacteria while in phase two there is a reproductive cycle during which predation does not occur. In phase one, the cell locates the prey by a process of "chance" collision followed by attachment and entry into the host cell. Once entrance has been achieved, the predating cell will reproduce feeding off the cells contents. In phase two, there is a shortage of possible target cells for predation and so reproduction is by the formation of long helical cells which then divide to form the typical polar flagellated cell.

Vampirovibrio

These bacteria are found in eutrophic (polluted) aerobic fresh waters. These cells have the ability to attach to the cells of some of the micro-algae such as *Chlorella.* Once attached, the cell becomes a "leech-like" parasite utilizing the excreted metabolic products of the host cells. These bacteria (like *Bdellovibrio*) can actually penetrate the cell walls of the cells being parasitized. These cells are comma-shaped, gRAM negative cells that are motile by a single unsheathed polar flagellum.

Helicobacter

A relatively new genus, *Helicobacter,* was first described as a cause of gastritis and duodenal ulcers in humans by Marshall and Warren in 1983. It is a gRAM negative curved or spirally shaped polar flagellated motile bacteria (Figure 8.6) that is microaerobic, both catalase and oxidase positive and generates a very strong urease activity. Motility is of a "darting" type. *H. pylori* is the recognized agent causing gastritis, duodenal ulcers that can then lead to a generalized ulceration of the gastro-intestinal tract.

Section 3, Non-Motile Straight and Curved gRAM Negative Rods

Spirosoma

These have been widely isolated from soils and freshwater. They are non-motile straight to curved rods which may form into long sinuous filaments of up to 50 microns in length. All are strict aerobes and are catalase and oxidase positive. They are only able to use oxygen as the terminal electron acceptor. Acids are produced by the metabolism of a variety of carbohydrates.

Section 21, Prosthecate (Stalked) Bacteria

These bacteria all bear a common characteristic which is the ability to generate a tubular extension (prostheca, stalk) from the cell (Figure 8.7). In some cases, the nuclear material is able to pass through the extension while in other cases, the stalk may be false. In the latter case, this represents a deliberate extrusion of (waste) materials from the cell itself rather than as a living extension of the cell itself. There are three genera differentiated primarily on the form of this extension and the nature of the life cycle (Table 8.4).

Table 8.4
Differentiation of the Genera in
Section 21, Prosthecate (stalked) Bacteria

	Prost	Bud	holdfast	Efe
Hyphomicrobium	true	+	-	-
Caulobacter	true	-	+	-
Gallionella	false	-	-	+

Prost, form of the prostheca (stalk); true, the stalk can contain cellular material; false, stalk is inaminate; Bud, ability of the bacteria to generate buds on the stalk; holdfast, ability of the bacterial cell to stick to surfaces; Efe, ability of the bacteria to obtain energy from the oxidation of iron.

Hyphomicrobium

Species are found in a wide variety of natural environments from soils to both fresh- and groundwaters. This genus has been recovered from porous media such as clays and it has been observed that the growing stalk allows the bacteria to penetrate materials with very small pore diameters (e.g., 0.2 to 0.4 microns). They are gRAM negative aerobes unusual in that neither sugars nor amino acids can support growth. However, strains are normally able to utilize ethanol and acetate. Strains are O.J.-II and can flourish on single carbon compounds such as methanol, formaldehyde and formate. They are considered to be facultative methylotrophs which means able to utilize methane and/or methanol as well as other somewhat larger organic molecules. These bacteria can dominate in oligotrophic (nutrient poor) environments. There is an unusual life cycle in which the stalk can extend with a diameter of 0.2 to 0.3 microns into the surrounding medium and then form a bud at the far tip of the stalk. The nucleus may then divide and one of the product nuclei may now pass along the narrow stalk to form a bud-like extension which now functions as a vegetative cell. By this manner, hyphomicrobia can penetrate and pass through very tight porous structures (e.g., clays).

Caulobacter

These bacteria are often found in very nutrient-starved conditions such as would be encountered in distilled or deionized water. They are also commonly isolated from fresh- and marine- water environments. They will often attach to other microorganisms to absorb nutrients being released from these cells. In this genus, the stalk is always produced from the polar regions (ends) of the cell which is also normally polar flagellated and rod shaped. A holdfast is also commonly present at one of the polar regions. This holdfast allows the cell to attach, in a semi-permanent manner, to any solid surface (living or dead!). The nucleus is not able to pass into the stalk in this genus and the role of the stalk appears to be limited to sometimes supporting the holdfast and providing a greater surface area for the absorption of nutrients.

Gallionella

This very distinctive genus is normally found in iron-rich surface- and ground-waters growing at the redox front between oxidative and reductive conditions. Very often, *Gallionella* is associated with the aerobic interfaces of iron-related bacterial biofouling of various water systems. A recognized major bacteria involved in water well plugging, they are gRAM negative O.J.-I and O.J.-II and are short or curved rods which may or may not be polar flagellated (life cycle in Figure 8.8). Some energy is obtained from the oxidation of the ferrous ion (Fe^{2+}) to the ferric ion (Fe^{3+}) at neutral pH values. The ferric oxides and hydroxide products are excreted from the cell into a stalk formed by extracellular polymeric substances that have been generated as a series of strands which form into either a flat or spiral ribbon extruding from a lateral region of the cell.

Sections 23 and 24, Gliding Bacteria

A very unusual group of bacteria are the gliding bacteria. This is because they have the ability to move when in contact with a surface but without the use of flagella, fimbrae or other observable organelles (Figure 8.9). Consequently, the precise nature of the gliding mechanism remains undetermined. The genera are differentiated in Table 8.5 (Figure 8.10). On occasions, the cells are observed to flex or "twitch" or rotate around the axis of the cell during the gliding motion. In all probability there are a number of mechanisms that may be employed by different strains of gliding bacteria. Movement can, however, be rapid with observed speeds reaching between 150 to 600 microns per minute.

Table 8.5

Differentiation of Gliding Bacteria in Sections 23 and 4

	Bacl	Cely	H_2SrS	Microc	Cella
Myxococcus	+	-	-	+	IR
Cytophaga	-	+	-	-	IR
Beggiatoa	-	-	+	-	FC

Bacl, bacteriolytic; Cely, cellulolytic; H_2SrS, hydrogen sulfide oxidized to sulfur; Microc, microcysts; cella, cell arrangement; IR, independent rod; FC, filamentous chains.

Myxococcus

These bacteria are found growing in slime formations moving over bacterially rich substrates such as cattle and rabbit feces. Their major nutrient source is the bacteria in the dung. They are all gRAM negative, aerobic, slender cells which may have slightly tapering ends or be flexible with rounded ends. As the cells mature, they tend to become coccoid or ellipsoid in form and highly refractile (microcyst). All are bacteriolytic predating mainly upon gRAM negative bacteria such as *Spirillum* and *Escherichia coli*. None are cellulolytic. All have the ability to strongly absorb congo red. The life cycle is simple and the microcysts are not enclosed in any form of sporangium. O.J.-III requirements means that strains usually require several amino acids for growth. Some strains are protease and/or lipase positive. Cells commonly occur within a common slime matrix (slime bacteria).

Cytophaga

Another group of slime bacteria, bacteria belonging to this genus are usually able to digest complex polysaccharides and all are able to degrade cellulose. Many, particularly marine and soil forms, are also often able to attack chitin, keratin and/or pectin. Some strains will even attack agar (agarolytic). This genus also plays a major role in mineralization processes and also in the decay of wooden structures. Strains have also been found to be major components in the treatment of sewage. *C. columnaris* is pathogenic to some fish causing a variety of conditions ranging from fin rot to columnaris diseases. They are free-living aerobic forms which do not generate a slime matrix. Cells are gRAM negative, flexible rods with pointed ends.

Beggiatoa

These are widely distributed in many habitats associated with soil and water. In particular, they occur at the redox fronts in aquatic systems particularly where H_2S is present on the reduced side of the front. Strictly aerobic or microaerophilic, the cells form into chains within colorless filaments. These filaments are flexible and are capable of movement through a gliding action. They lack a sheath but other closely related genera do bear a sheath around the filament(s). Metabolically very versatile, H_2S can be oxidized leading to the deposition of large sulfur granules within pockets formed in the cell structures. Subsequently, these sulfur granules can be oxidized further to sulfate. In this process, energy is generated through the electron transport chain using the sulfur electrons. Growth can also occur using acetate as the sole carbon source.

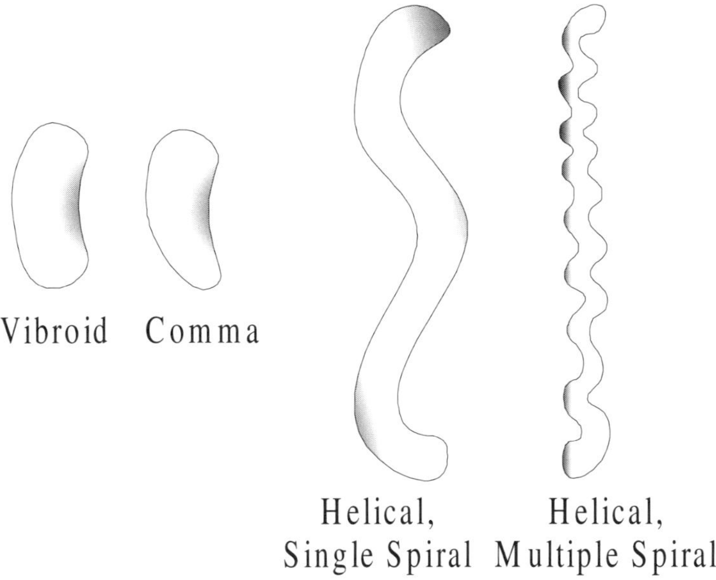

Figure 8.1 Cell forms observed in the Section 1, 2, and 3 gRAM negative helical and spiral bacteria.

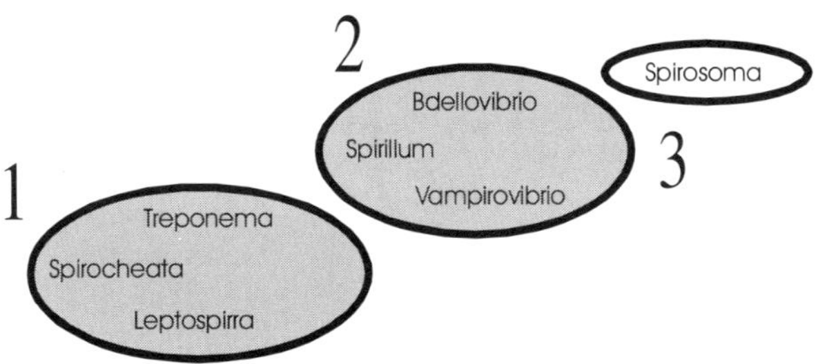

Figure 8.2 Differentiation of Sections 1, 2, and 3 gRAM negative curved and spiral bacteria by motility (shaded).

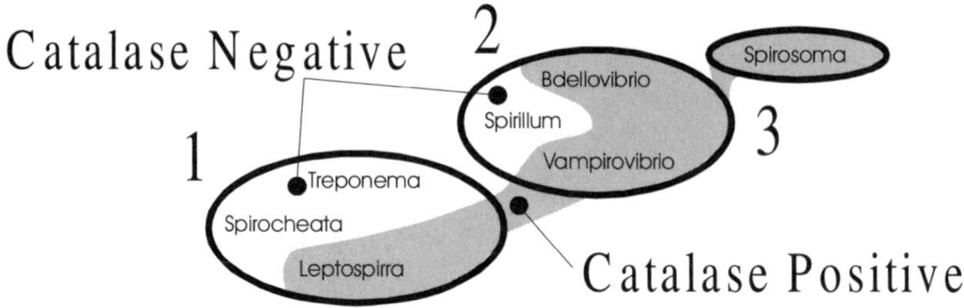

Figure 8.3 Differentiation of Sections 1, 2, and 3 gRAM negative curved and spiral bacteria by the presence or absence of the catalase enzyme.

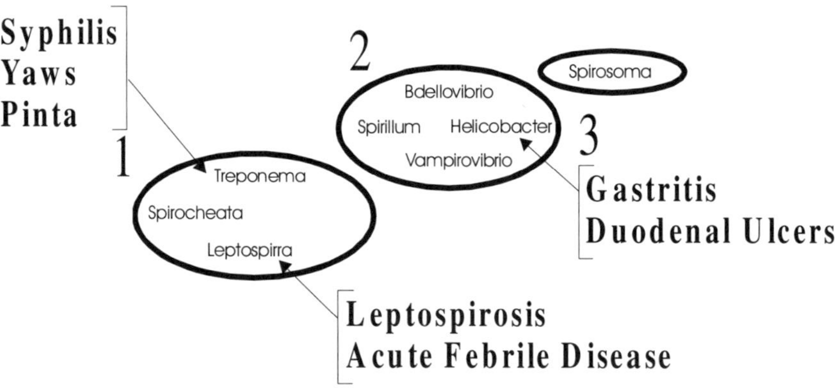

Figure 8.4 Diseases in humans caused by members of Sections 1, 2, and 3. A separate bracket is displayed for each genus known to be pathogenic to humans. An arrow with a continuous line connects the bracket to the name of the genus responsible for those diseases.

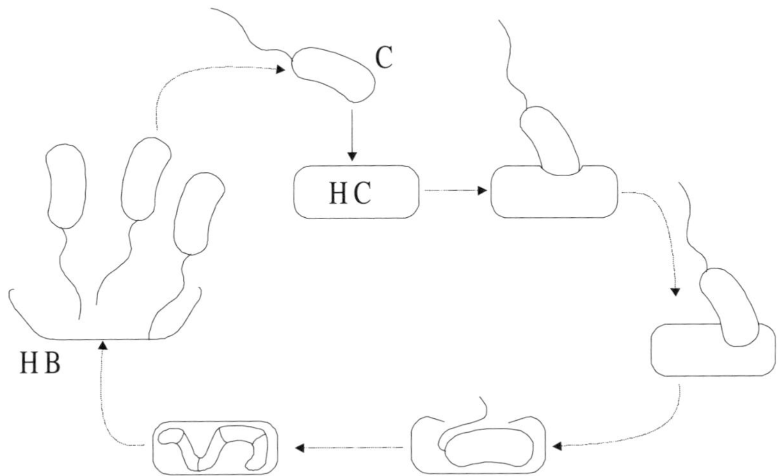

Figure 8.5 Diagrammatic representation of the life cycle of the Section 2 bacteria, *Bdellovibrio*. Cell (C) enters host cell (HC); grows and reproduces causing the host cell to break apart (HB); releasing the daughter cells to continue the infectious process.

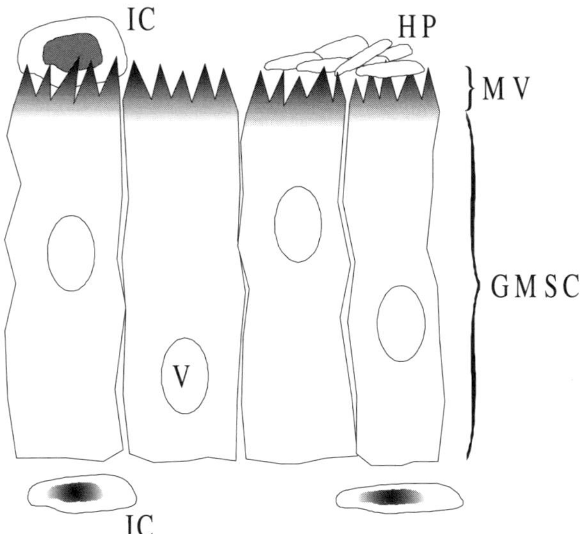

Figure 8.6 Colonization of the mucous layers in the gastro-intestinal tract by *Helicobacter pylori* (HP), a major cause of duodenal ulcers and gastritis. The cells grow over the microvilli (MV) in the vacuolated (V) gastric mucous-secreting epithelial cells (GMSC). The region is patrolled by inflammatory cells (IC) called neutrophils.

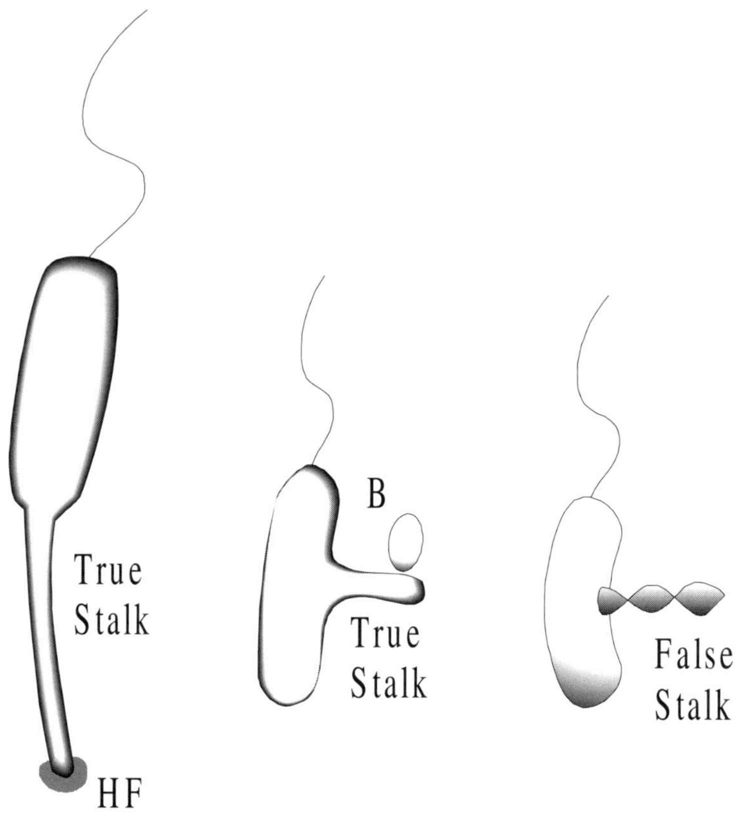

Figure 8.7 Illustration of the principal defining differences between the three genera of stalked (prosthecate) bacteria based upon the nature of the stalk and the ability to produce buds (B) and holdfasts (HF).

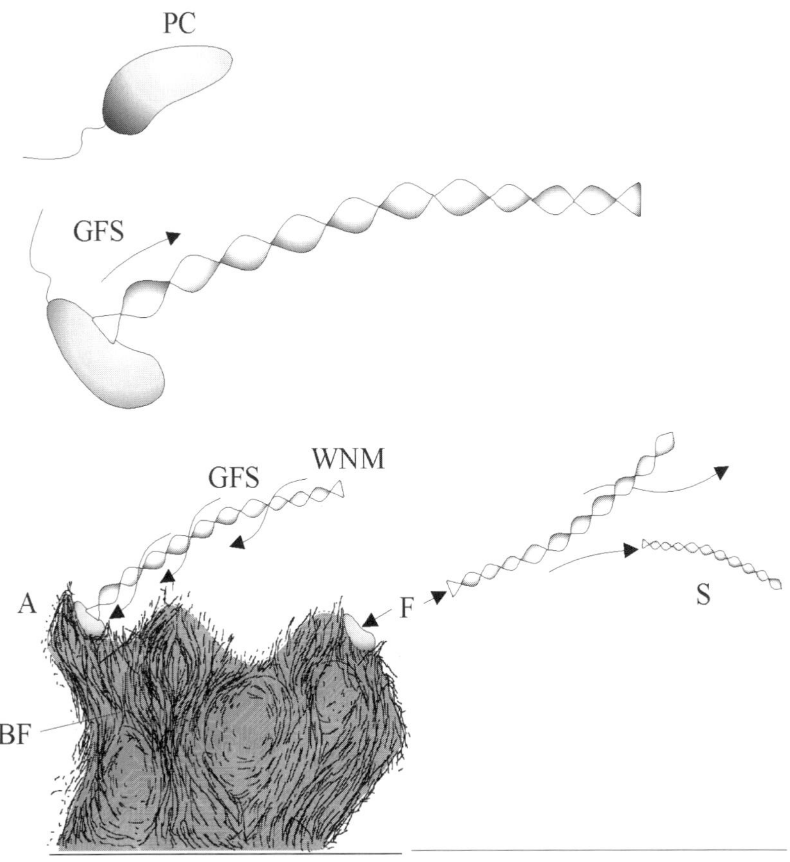

Figure 8.8 Diagrammatic representation of the life cycle of *Gallionella*. A free swimming planktonic cell (PC) attaches (A) to the surface of the biofilm (BF) and generates a ribbon-like false stalk (GFS). When the stalk gets too long, it fractures (F) and the stalk (S) floats away.

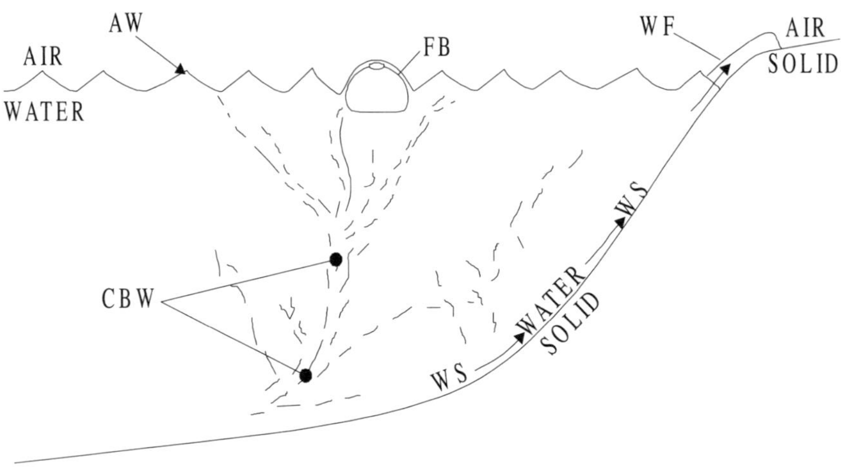

Figure 8.9 Representation of the forms of gliding that are observed in Sections 23 and 24, the Gliding Bacteria. These bacteria may move along the air:water (AW) or water:solids (WS) interface within films forming bubbles (FB), inside of colloidal bound water (CBW) structures within the water column, or on water films (WF) in unsaturated porous media.

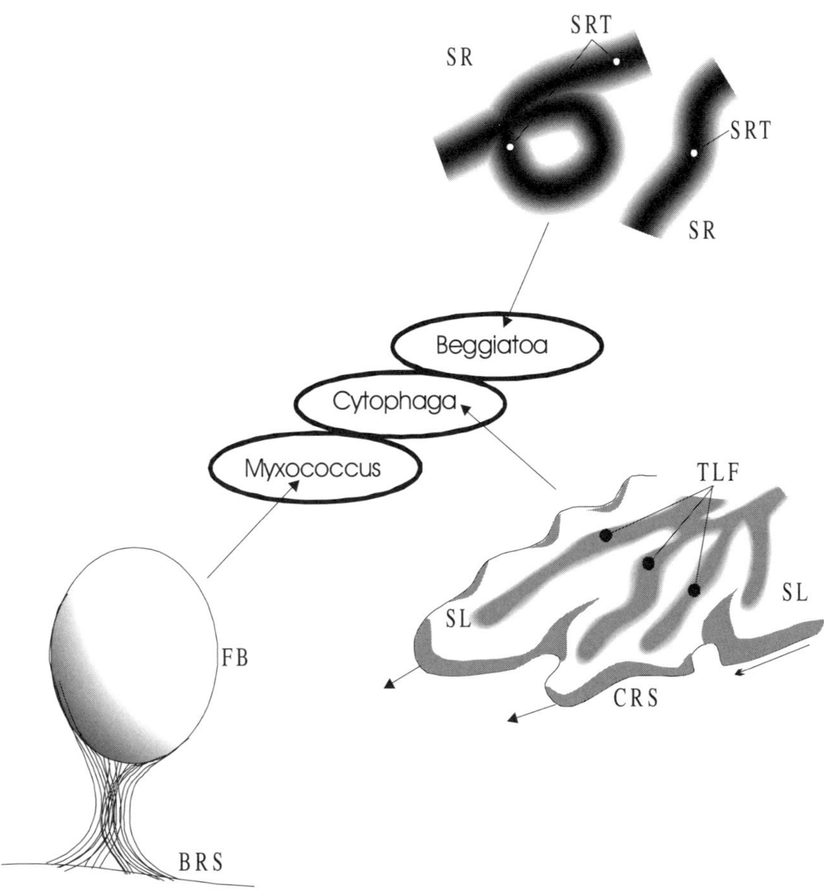

Figure 8.10 Illustration of the comparative cell forms for the gliding bacteria, *Myxococcus, Cytophaga,* and *Beggiotoa. Myxococcus* commonly generates fruiting bodies (FB) that grow over a bacterially rich substrate (BRS) such as dung or fecal pellets. *Cytophaga* grows as a flowing slime (SL) in which the cells assume a thread-like form (TLF) and grow over the surfaces of a cellulose rich substrate (CRS) such as a rotting tree trunk. *Beggiotoa* has sulfur granule rich trichomes (SRT) that resemble a filamentous mass and grows in aquatic hydrogen sulfide rich (SR) environments.

9

Consortial Bacterial Forms

In the environment, it is very unusual to find a single strain of bacteria growing in a unique site without direct interaction with other strains of microorganisms. In reality, it is very common to find a number of microbial species co-existing within a community. This community is often referred to as a "consortium." While in infectious diseases, it is common to focus in on a single strain that is believed to be the pathogenic agent, microbial ecologists have to focus in on the consortia of bacteria that are "infesting" the site being examined.

This chapter addresses the need for, and ability to achieve, an identification system for bacterial consortia. To date, the major method to identify the consortium has been to examine the form and nature of the growth and the environment within which it is growing. Examinations have commonly been concentrated on the isolation and identification of the dominant bacterial strains grown using the agar plate or the membrane filtration techniques. This does not give an overview of all of the component strains that may be active within the consortium, or their relative aggressivity. The term "aggressivity" reflects the potential activity level of the strains rather than simply their cell numbers and relative populations.

A consortium (Figure 9.1) may be simply defined as a group of bacterial strains that are able, through being interdependent upon each other of the strains' abilities, to form a coherent structure within, and upon which all of the strains' coexist. These consortia may take many forms ranging from floating (biocolloidal) suspended solids through various slime thread and coatings to crystallized structures such as nodules, encrustations, rusticles, iron pans, plugs, plugging wells, stalactites and stalagmites, and concretions. In the past, it was common for many of these consortial structures to be considered undefinable chemical events involving complex chemistries. Today, the improvements in biotechnology have led to improved abilities to detect the various bacteria in these consortia. The whole process of identifying consortia remains in its early phases and this chapter will concentrate on the application of the patented Biological Activity Reaction Tests (BART™, Droycon Bioconcepts Inc., Canada) to culture a range of these consortia.

The BART™ biodetector system (Figure 9.2) relies on the generation of gradients for the movement of nutrients and reductive conditions upwards within a 15ml liquid sample. Nutrients that are selective for the specific group of bacterial consortium are presented as a crystallized pellet on the floor of the test device. The oxidative - reductive (redox) front is created by a floating ball that restricts oxygen entry into the liquid medium. Incubation is commonly and conveniently at room temperature. This temperature clearly supports the "low" mesophiles primarily and these are the bacteria most often associated with near-surface microbial activities. Samples taken from colder or warmer regimes would require a different temperature. For a sample taken from cold environments (i.e., <15°C), a good incubation temperature would be 8°C, a typical temperature in some of the "warmer" refrigerators. For higher temperatures, the incubation temperature should be within 10°C of the original sample temperature. For isolates from, or related to, warm blooded animals, a standard incubation temperature would be in the 35 to 37°C range. This is often referred to as "blood heat."

Typical sample innocula range from 15ml for a water sample to the supplementation of a 15ml volume of sterile phosphate buffer with the sample. Commonly, 0.1g is used for a soil sample, 0.1ml for a dispersed material (such as slime) or one loop-full of a broth culture. For the soil sample, it is advised that the soil should be added to the test vial with the sterile ball aseptically removed immediately prior to adding the soil sample and returned immediately afterwards and before the 15ml of diluent is added.

Observation of the BART™ test vials should be daily and the following events recorded:

1. The time lag (TL) to the first observation of a reaction attributable to the activities of the bacteria in the sample and not to the normal chemical processes associated with the diffusion of the media up through the redox front. This time lag can be used to indicate the aggressivity of the bacteria in the sample. Thus, the quicker the reaction occurs, the more aggressive the population.
2. The reactions created by the bacteria within the test vial occur over a sequence of time after the first reaction is recognized (at the TL). The reactions are all encoded as two letter summaries of the reaction observed. For example, BR would be the code for brown ring around the ball. The sequence in which these reactions are observed can be used to predict the range of species present within the consortium. This is known as the reaction pattern signature (RPS) and forms the basis for the identification of the consortia described below.

Common terminal (concluding) reactions for the BART™ tests described in Figure A.4 are: IRB, BR with either BL or BC; SRB, BA;

SLYM, either CL or BL commonly without an obvious slime ring. Tests may be considered completed when the these reactions are observed.

Identification of Consortial Components using the RPS

In general, all reactions (i.e., the completed RPS) being generated by the BART™ are over by day twelve and any indication of aggressive bacteria is over by day eight. Once the test has been completed, the order in which the reactions have occurred to form the RPS can be used to identify the potential major bacterial genera that would be present in the sample. The RPS remains subject to change (through additional observed reaction types) until the test is over unless a typical terminal reaction is observed. Each of the different types of BART™ biodetectors have been used to determine consortial RPS. Some of the consortia of mixed genera that are commonly seen for each of the defined RPS are listed below. It should be remembered that the application of this test system to determine consortial components is still new and will, undoubtedly, be subjected to changes as the data base expands. For each BART™ type, the recognized RPS is listed below. Individual reaction codes are listed and defined in the Appendix.

IRB-BART™

The following is a list (in descending order of occurrence) of the RPS that has been observed and each is followed by a list of the genera commonly associated with that particular RPS.

<u>FO-CL-BC-BR</u>, mixed community of Sections 4 and 5 along with some Section 12. Slime-forming and iron-related bacteria will be present. Genera most likely to be present would include: *Enterobacter, Serratia, Pseudomonas, Micrococcus, Gallionella, Bacillus* and *Alcaligenes*.

<u>CL-BC-BR</u>, aerobic iron-related bacteria are present in abundance along with slime-forming bacteria. This suggests that aerobic bacteria will be dominating with the following genera commonly present: *Pseudomonas, Alcaligenes, Micrococcus, Gallionella, Crenothrix, Bacillus* and *Enterobacter*.

<u>GC</u>, this RPS includes only one reaction due to the dominance of Section 4 bacteria. Where the reaction remains a moderately clear light green, the dominant genus is likely to be *Acinetobacter*. If the green goes darker and becomes cloudier, then the dominant genera would include *Pseudomonas, Zoogloea, Micrococcus, Alcaligenes* and *Janthinobacterium.*

<u>FO-GC-BL</u>, this reaction occurs where a sample has come from close to a redox front with a mixture of Section 4 and 5 bacteria. The FO as the first signal indicates the presence of anaerobic bacteria producing gases while the GC indicates that there is an aggressive population of species of

Pseudomonas, Acinetobacter and *Janthinobacterium* from Section 4 but Section 5 species then dominate with a mixed flora including *Enterobacter, Aeromonas, Serratia, Klebsiella, Aeromonas, Vibrio,* and *Proteus.*
CL-BG, this RPS may be preceded by an FO but the development of clouding followed by a brown gel (BG) is very typical of samples dominated by species belonging to the genus, *Enterobacter.*
FO-CL-RC, this is a common reaction where the sample is dominated with a range of enteric, Section 5 bacteria and it is usually species of *Enterobacter, Citrobacter, Klebsiella,* and *Serratia* that are dominant. If the RPS is followed by the BR (brown ring) it has been noted that the ring may occlude the ball (i.e., form a "glue") so that, if the BART™ test vial is turned upside down, the liquid remains perched above the ball that is glued to the walls of the test vial. This is a very common occurrence where strains of *Citrobacter freundii* dominate.

SRB-BART™

The following is a list (in descending order of occurrence) of the RPS that has been observed and each is followed by a list of the genera commonly associated with that particular RPS. The unusual feature in this test is that the SRB (commonly species of *Desulfovibrio*) are influenced (and protected) by other bacteria that are often capable of aerobic growth. There are four RPS recognized and three include the presence of *Desulfovibrio* (i.e., those that give a black, B, reaction). The fourth does not involve the activities of SRB.
BB, represents a mixed consortium of facultatively and strictly anaerobic bacterial species growing with the SRB. These additional species could include *Enterobacter, Aeromonas, Klebsiella, Leuconostoc, Proteus, Aeromonas, Chromobacterium, Lactobacillus* and *Actinomyces.*
BT, represents very aggressive aerobic bacterial populations within which the SRB form a part of the consortium. With this RPS, the SRB are integrated into an aerobic consortium that includes some slime-forming bacteria. The reaction itself often consists of little black specks (SRB growths) within a slime film of varying thicknesses growing over the submerged surfaces of the ball. The consortium here would be comprised predominantly of aerobic bacteria and would be likely to include: *Pseudomonas, Alcaligenes, Janthinobacterium, Acinetobacter, Vibrio, Micrococcus, Bacillus, Streptomyces* and *Klebsiella.*
BB-BA, this RPS begins with a basal blackening followed by a blackening of the ball. There is, in these cases, a much more aggressive consortium that is able to create very reductive conditions in which the SRB are now able to form a part of the communities growing around the ball. In addition to the

likely genera, *Enterobacter, Aeromonas, Klebsiella, Leuconostoc, Proteus, Aeromonas, Chromobacterium, Lactobacillus* and *Actinomyces*, there may also be some of the strictly aerobic genera that are able to cause a radical shift in the redox front around the ball. These could include *Pseudomonas, Micrococcus, Alcaligenes, Janthinobacterium, Bacillus,* and *Streptomyces.*
BT-BA, this RPS represents an extension of the aerobic consortium that originally is supporting the growth of the SRB around the ball. It would be the presence of facultatively anaerobic genera that would cause the extension of SRB activity down into the more reductive zones. In addition to the genera found in a BT reaction (*Pseudomonas, Alcaligenes, Janthinobacterium, Acinetobacter, Vibrio, Micrococcus, Bacillus,* and *Streptomyces*), there would also be more activity from additional members in the consortium that are facultatively anaerobic. These could include *Enterobacter, Aeromonas, Klebsiella, Leuconostoc, Proteus, Aeromonas, Chromobacterium, Lactobacillus* and *Actinomyces.*
CG-, represents the occurrence of a clouded gel-like mass in the liquid medium often before the onset of any positive SRB reactions. The bacteria causing this reaction usually generate a very reductive thick gel-like mass that could include: *Aeromonas, Leptotrichia, Bacteroides, Bacillus, Propionibacterium, Lactobacillus, Enterobacter, Citrobacter, Serratia,* and *Klebsiella.* Generally, the CG reaction precedes the detection of SRB presence (i.e., BB, BT and BA) and is not a positive indication of a SRB.

SLYM-BART™

The following is a list (in descending order of occurrence) of the RPS that has been observed and each is followed by a list of the genera commonly associated with that particular RPS. The observations of reactions in this test system revolve around, primarily, the generation of clouding. Typical RPS are described below.
CP-CL, starts with a stratification of cloudy plates usually at the redox front. These generally begin as a single plate that divides and finally disperses as a clouded growth. The dominant genus associated with this event is *Proteus* but other genera that may be present include *Micrococcus, Klebsiella, Pseudomonas, Lactobacillus* and *Chromobacterium.* These other genera dominate later.
DS-CL, this RPS reflects a dominance of slime-forming bacteria that produce a deep gel-like growth initially followed by a generalized clouding. Genera associated with consortia causing this RPS can include *Alcaligenes, Klebsiella, Enterobacter* and *Flavobacterium.*
SR-CL, is a rarer RPS in which the initial slime ring is dominated by the activities of *Micrococcus* associated with *Pseudomonas, Zoogloea,* and

Alcaligenes.
<u>Cl-BL</u>, is a common RPS where there is a mixed consortium in which the strict aerobes (*Pseudomonas, Alcaligenes, Janthinobacterium, Acinetobacter, Vibrio, Micrococcus, Bacillus,* and *Streptomyces*) dominate at the beginning but are supplanted by enteric bacterially dominated consortium (*Enterobacter, Aeromonas, Klebsiella, Leuconostoc, Proteus, Aeromonas, Chromobacterium, Lactobacillus* and *Actinomyces*).
<u>CL-PB</u>, is an RPS that occurs when *Pseudomonas aeruginosa* is sufficiently aggressive that it dominates on the oxidative side of the redox front and generates a pale blue flourescent pigment that glows in U.V. light.
<u>CL-GY</u>, is an RPS that occurs when *Pseudomonas fluorescens* is sufficiently aggressive that it dominates on the oxidative side of the redox front and generates a greenish-yellow flourescent pigment that glows in U.V. light.

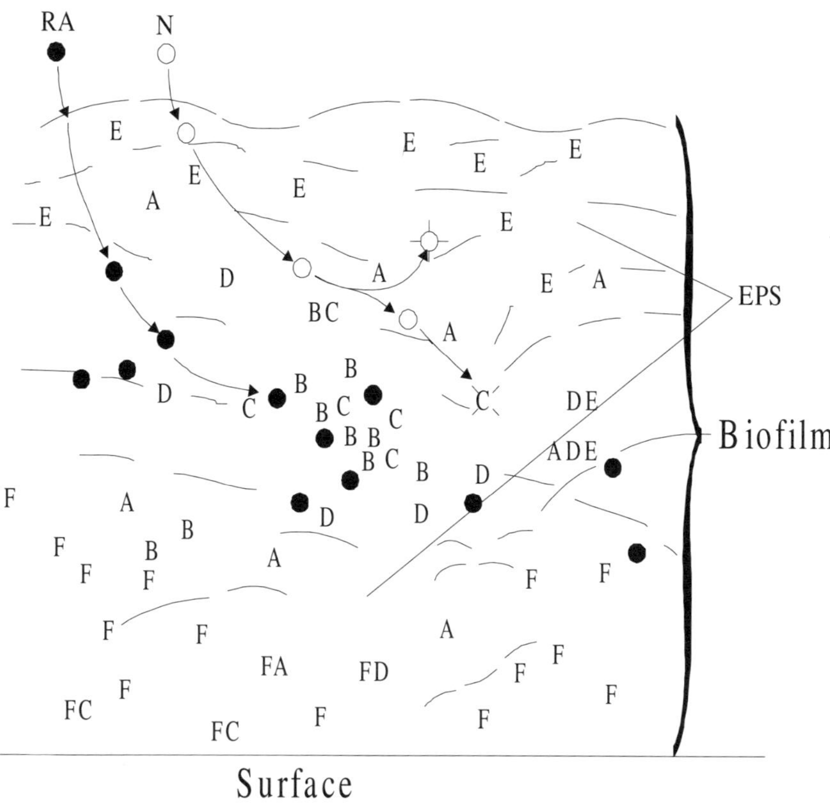

Figure 9.1 Diagram of the manner in which various species can form into a consortium within a biofilm. The species are labelled A, B, C, D, E, and F and are dispersed within the extracellular polymeric substances (EPS) commonly referred to as "slime." The EPS contains racalcitrant accumulates (RA, closed circles) such as ferric iron and also nutrients (N, open circles) that move through the EPS to be utilized by the cells. Assimilatable nutrients will move towards (arrows) active cells in the consortium while the RA will continue to passively accumulate.

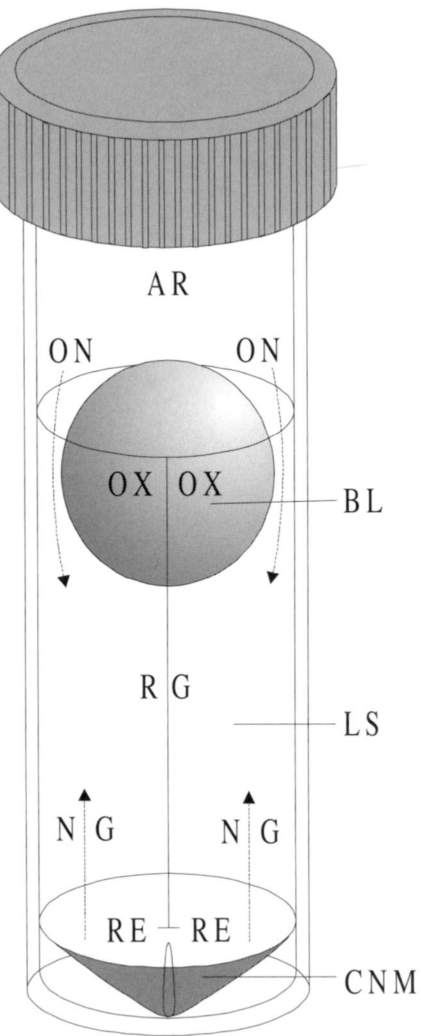

Figure 9.2 Diagrammatic representation of the functional operation of the BART™ test. Two vertical gradients are created under, and around, the ball (BL) floating on 15 ml of liquid sample (LS). Oxygen (ON) diffuses down from the air (AR) in the head space above the ball. This creates an oxidative zone (OX) above the reductive zone (RE) in which any residual oxygen is rapidly removed. A redox gradient (RG) therefore forms down the length of the liquid. At the same time, the crystallized nutrient medium (CNM) applied inside the base begins to dissolve upwards to form a nutrient gradient (NG). The two gradients are therefore the redox gradient (RG) and the nutrient gradient (NG). Bacterial growth therefore tends to focus at the most suitable position(s) along the NG and RG.

10

The Atlas Concept

The bacterial kingdom is, by its very nature, a very large and diverse collection of microorganisms that can only be easily appreciated by their activities and, more occasionally, the form and size of the structures that they sometimes build. Over the last century since the "golden age" of bacteriology (1878 to 1888), there has been an emphasis on the bacteria that are potentially pathogenic to humans or the animals of economic consequence. This not unnatural emphasis has led to a much greater concentration of effort to determine the nature of pathogenic bacteria rather than those that may have environmental consequence. As funding and needs followed this trail towards understanding potentially pathogenic bacteria, so the whole field of environmental microbiology was ignored or simply reduced to a series of "natural" chemical reactions that were abiotic. The use of the word "abiotic" in this concept is interesting because, by the dictionary definition, it would mean no biota were present. Biota may be defined as plants and animals and so the term abiotic can be interpreted to mean without life (i.e., sterile) while in reality it simply means no plants or animals were present. The net impact of this covert belief is that there has been a devaluation of the activities of bacteria beyond that associated with pathological conditions.

An example of this focussing was soil bacteriology. Soil contains a rich diversity of bacteria commonly dominated by the mycelial bacteria (Sections 26, 27 and 29) with fluctuating seasonal growths of cyanobacteria (these used to be known as the blue-green algae). During the latter part of the century, the most attention has been paid to the genus *Rhizobium,* because of its role in the nodulation and fixation of dinitrogen (N_2) in leguminous crops that form a major part of the economic and value of the agronomic crop rotation cycle.

The neglect in understanding many parts of the bacterial kingdom has led to biases favoring the identification of bacteria known to be of medical and veterinary significance. In the last thirty years, the improvements in the chemical analysis of nucleic materials have opened up new approaches to the classification of bacteria but the Achilles' heel for this process is that bacteria, unlike other major biological groups, commonly grow in communities (consortia) in which there is an interdependence between the component species. The myopia of identification today is that classification has to be performed on a single strain of bacteria (and not a consortial mix). In the processes of purifying a single strain for identification, many of the strains within the consortium are lost even if they were actually major contributors. Emphasis favors those that are easily culturable. Selecting the conveniently culturable bacteria does not necessarily give a picture representative of all of the bacterial biomass involved in the sample under investigation.

This atlas chapter has been designed using the conventional Bergey's Manual rather than a blended combination of the more recent classification systems using the newer molecular technologies. The reason for doing this is that this is an introduction to the classification of bacteria using the traditional approaches. The atlas is a format that was used in the previous chapters to define the interrelationships of the bacterial genera within sections and families. This book has been designed to introduce the reader to the concepts upon which the identification of bacteria has been evolved given the limitations inherent in ignoring the role of bacterial consortia in most, if not all, bacterial activities.

The format used for this atlas is to borrow the two dimensional concepts commonly used by geographers to map regions. Here, however, the "continents" are sections, the "countries" are families and the "cities" are genera. In developing the atlas, there were a number of basic considerations (Figure 10.1). These include the gRAM reaction moving from negative (Figure 10.2) to positive (Figure 10.3). Following this, the next two figures (10.4 and 10.5) delineate divisions A and C using the concepts employed in "The Prokaryotes." This is followed by aerobicity (Figures 10.6 through 10.8) and cell shape (Figures 10.9 through 10.11). In the next figures (10.12 and 10.13) the two main forms of motility are displayed involving polar and peritrichous flagella. Following this, the atlas addresses the genera possessing specific enzyme systems including catalase (Figure 10.14), oxidase (Figure 10.15), proteases (Figure 10.16), and lipases (Figure 10.17). Genera able to denitrify (nitrate being reduced to nitrite) are displayed in Figure 10.18. The next three atlas charts (Figures 10.19 through 10.21) show the range of bacterial genera able to produce pigments of specific colors such as yellows, reds, pinks and violets. Genera able to degrade glucose and lactose to acidic or gaseous products are addressed next in Figures 10.22 through 10.25. Figures 10.26 and 10.27 cover the ranges of bacterial genera able to parasitize other microorganisms and plants respectively. Bacterial spoilage of foodstuffs is a major economic as well as hygienic concern and Figures 10.28 through to 10.30 show which genera cause spoilage in dairy, poultry and meat products. This is followed by two atlas sheets showing the common genera found in soils (Figure 10.31) and waters (Figure 10.32). The next three figures (10.33 to 10.35) deal specifically with defining the bacterial genera that can cause diseases in humankind by indicating which genus is responsible for each of a list of common diseases presented in tabular form by genus. The remaining figures in this chapter are devoted to defining the generic composition of various bacterial consortia detectable using the BART test systems (Figures 10.36 through to 10.45).

Two ways in which the atlas can be used is to determine which bacteria are likely to be present given the environmental conditions at the site of concern, and confirmation of the characteristics of isolated bacteria reflecting the elected identification of the bacterial strain.

Consortial identification remains in its infancy and has been addressed to some extent in Chapter Nine. This is a very new area of bacteriology and the true nature, form and function of consortial activities remain embryonic in the environmental aspects of bacteriology and virtually unrecognized in medical and veterinary bacteriology. The potential role of consortia acting as relatively passive vectors for carrying a pathogen to an infection site, be it viral or bacterial, remains to be explored.

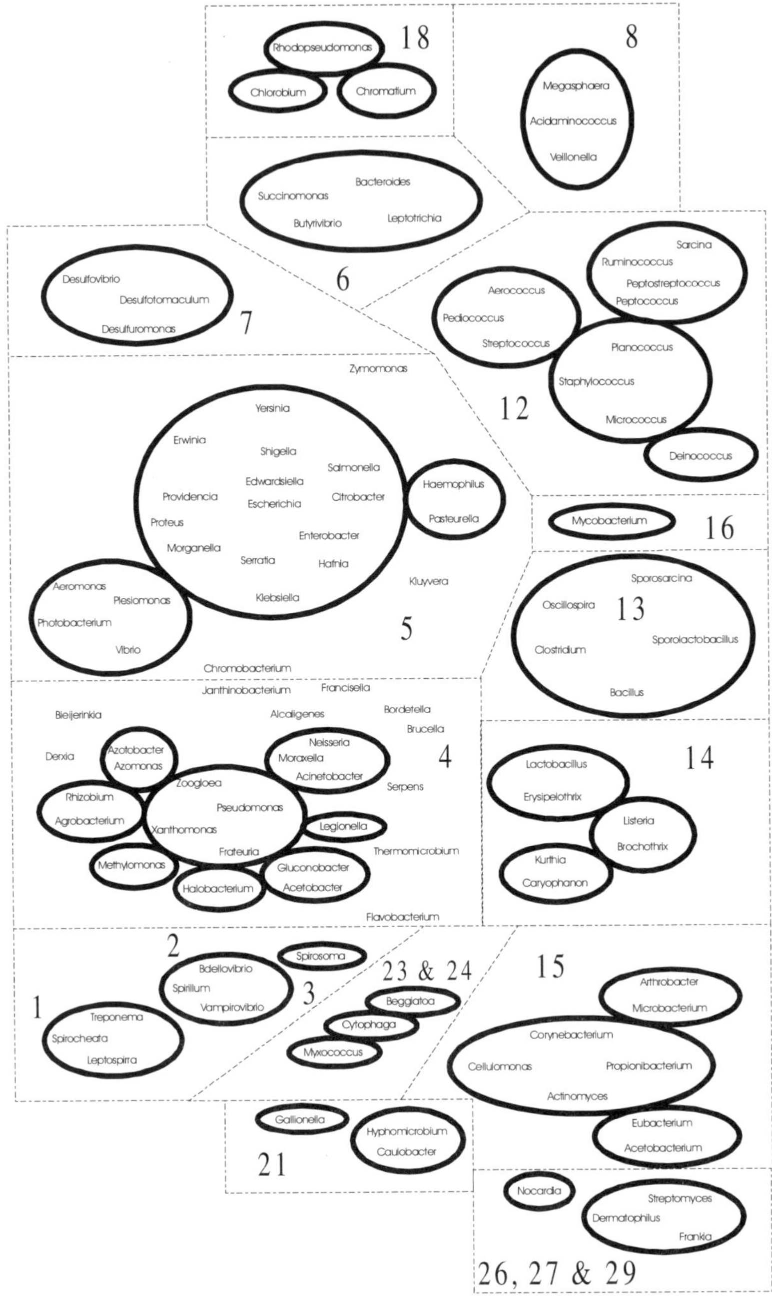

Figure 10.1 Differentiation of the bacterial genera into sections displaying the relative positions of the various sections (given as numbers within the boundaries). Note that these positions form the atlas concept and will be used in the subsequent figures in this chapter.

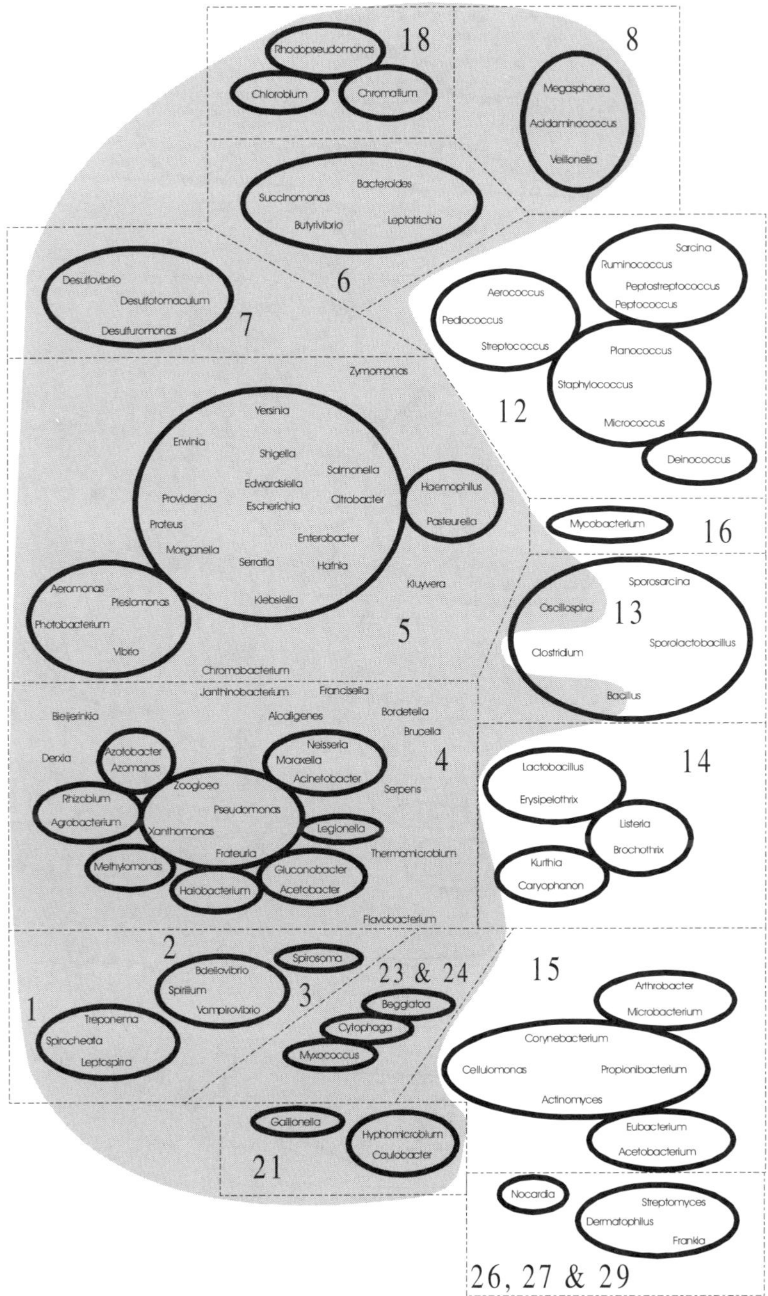

Figure 10.2 Atlas displaying as the shaded zones those bacterial genera that are commonly gRAM negative.

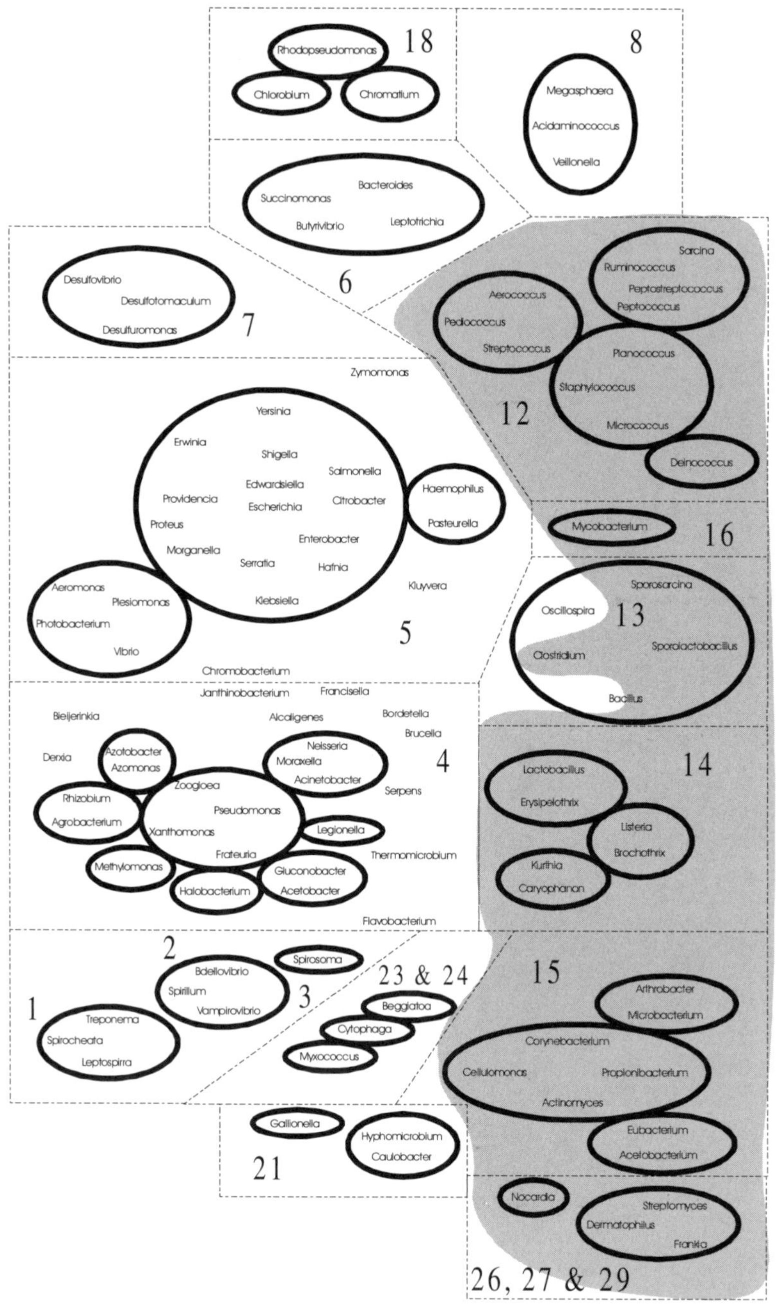

Figure 10.3 Atlas displaying as the shaded zones those bacterial genera that are commonly gRAM positive.

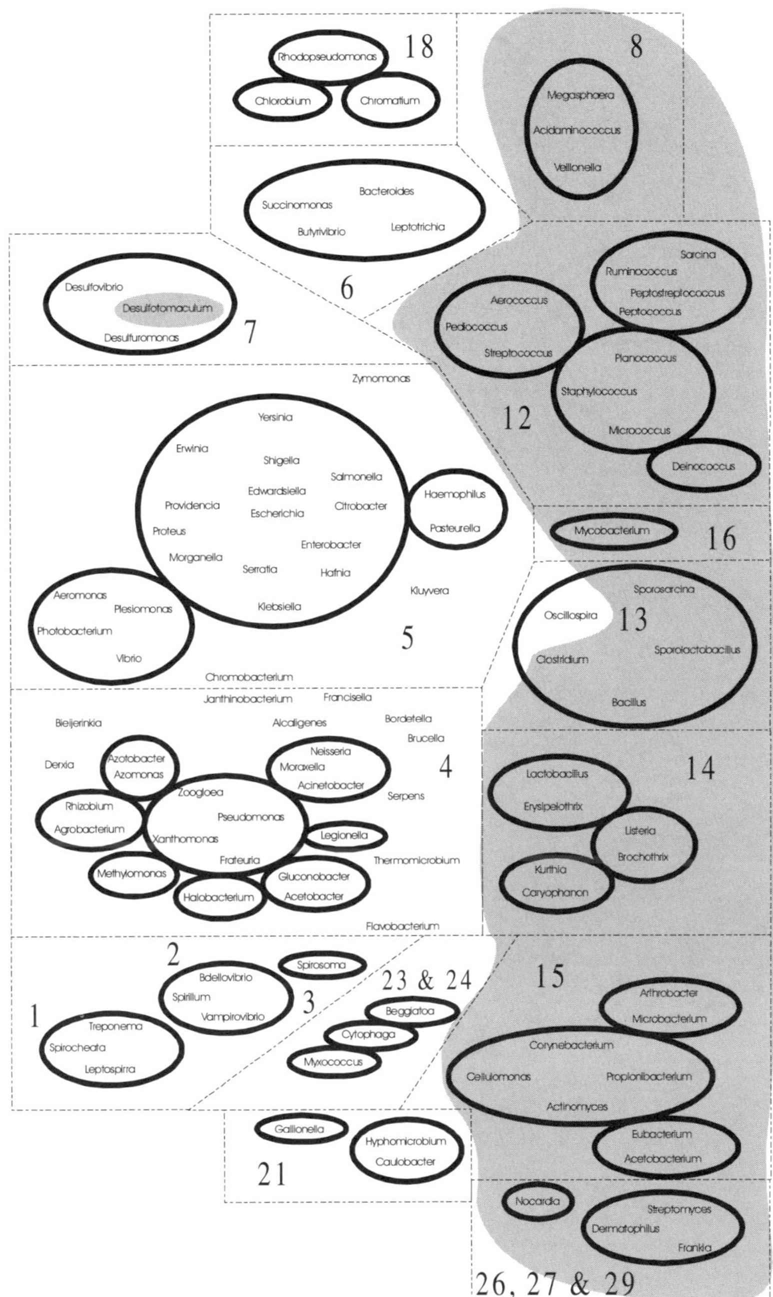

Figure 10.4 Atlas displaying as the shaded zones of those bacterial genera commonly associated with the Division A, *Firmicutes* (based on the second edition of The Prokaryotes)

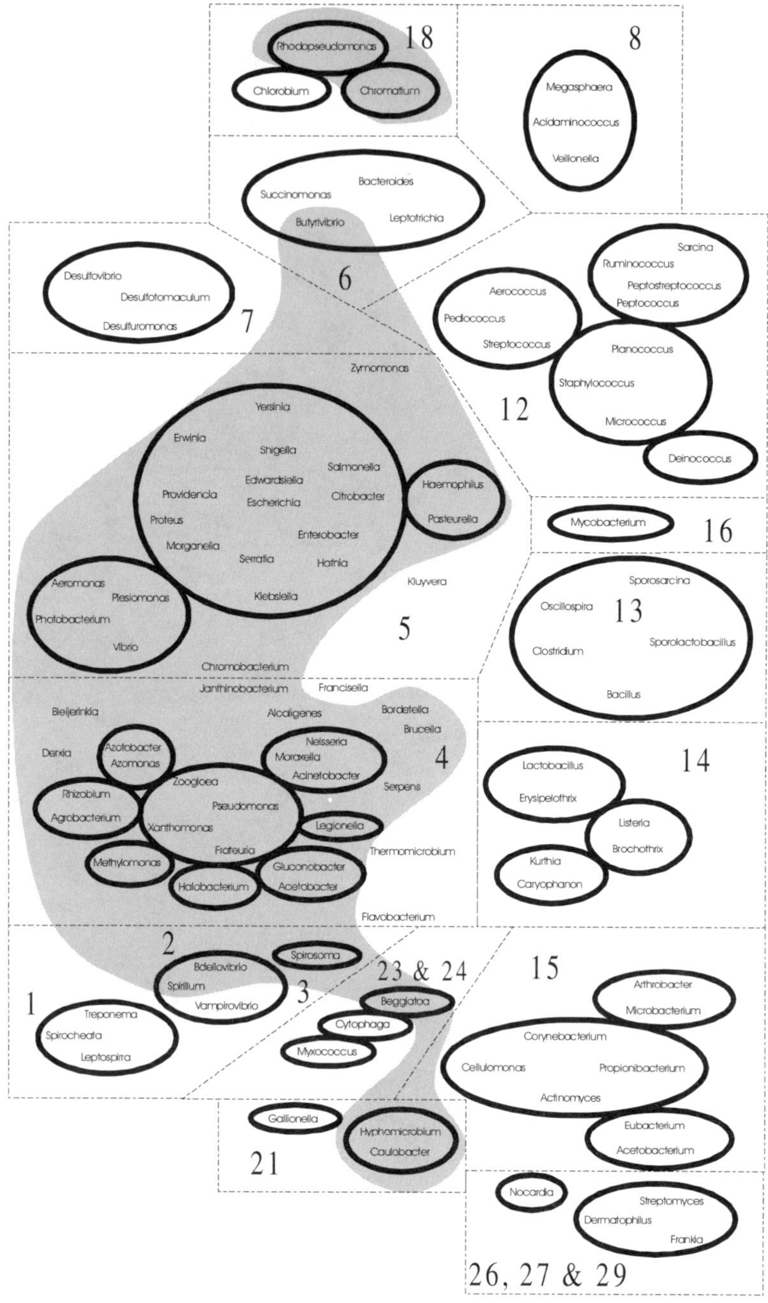

Figure 10.5 Atlas displaying as the shaded zones of those bacterial genera commonly associated with the Division C, *Proteobacteria* (based on the second edition of The Prokaryotes).

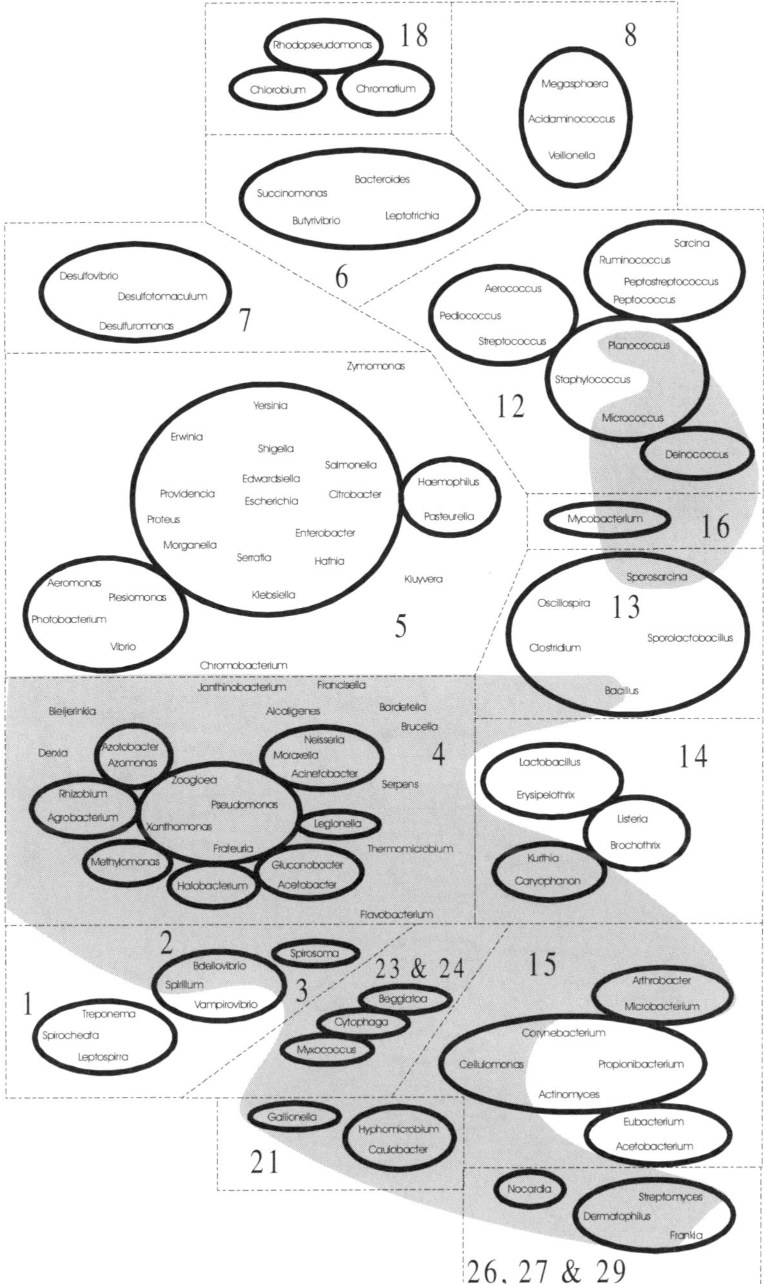

Figure 10.6 Atlas displaying as the shaded zones those bacterial genera that are commonly strictly aerobic.

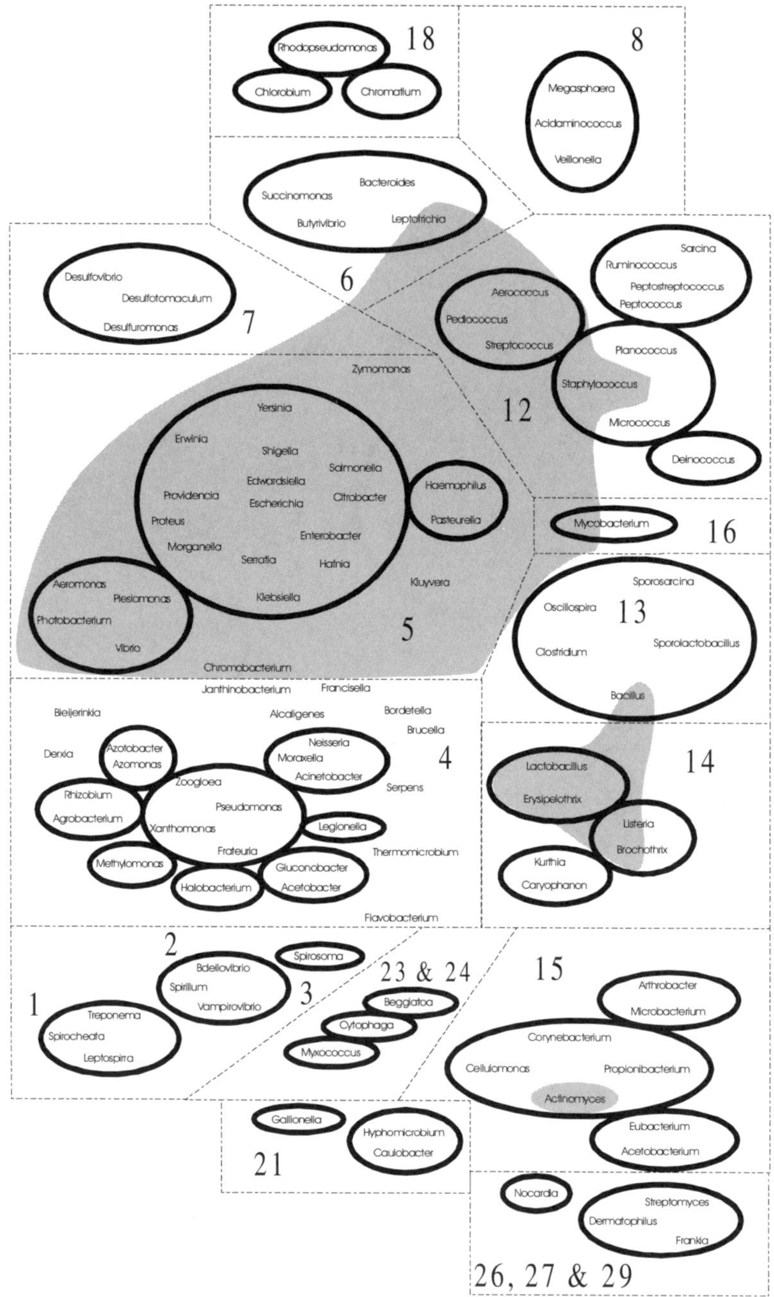

Figure 10.7 Atlas displaying as the shaded zones those bacterial genera that are commonly facultatively anaerobic.

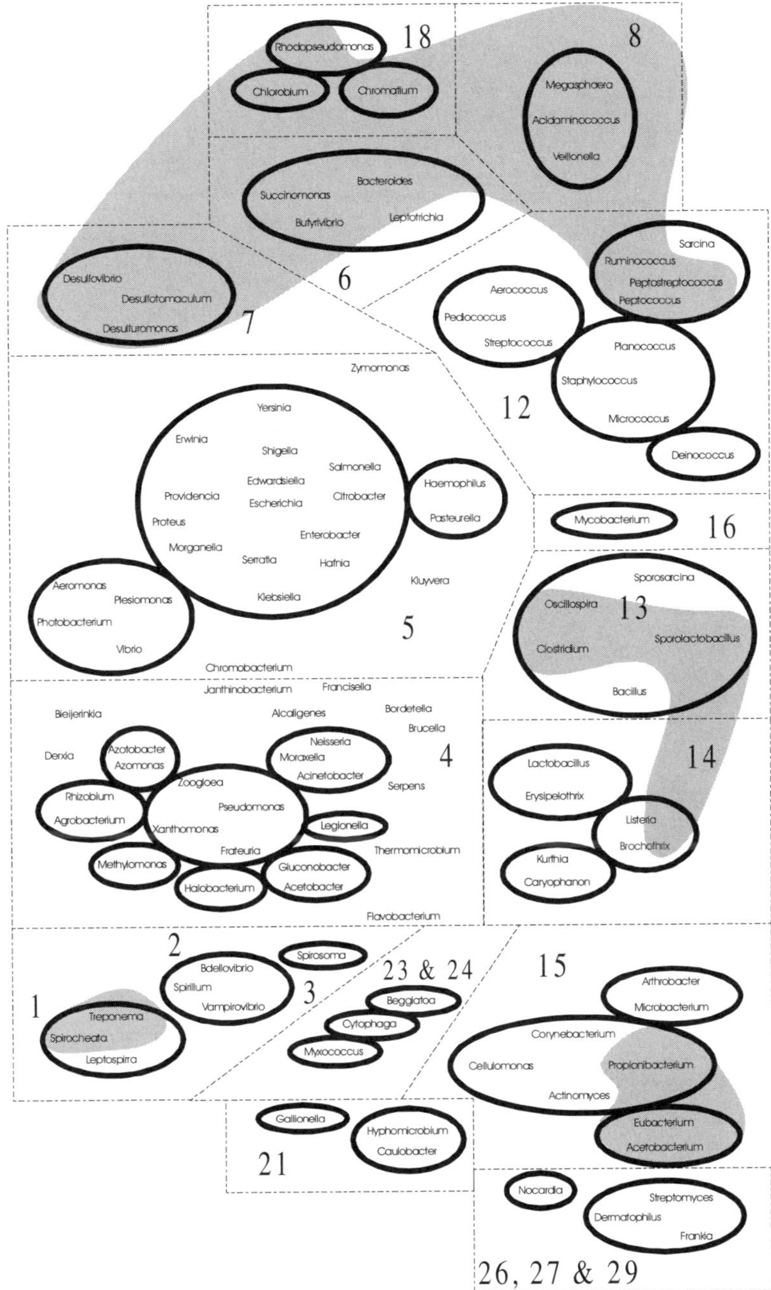

Figure 10.8 Atlas displaying as the shaded zones those bacterial genera that are commonly strictly anaerobic.

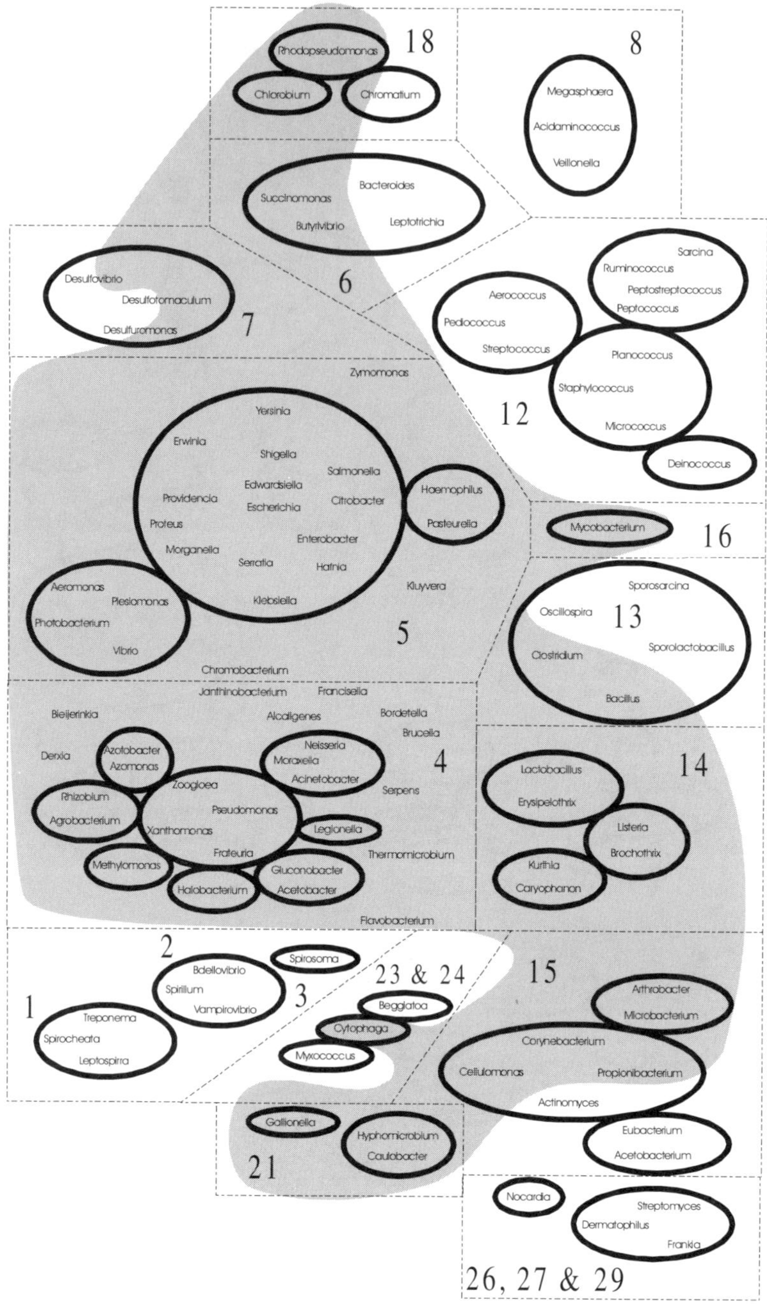

Figure 10.9 Atlas displaying as the shaded zones those bacterial genera that are commonly rod shaped.

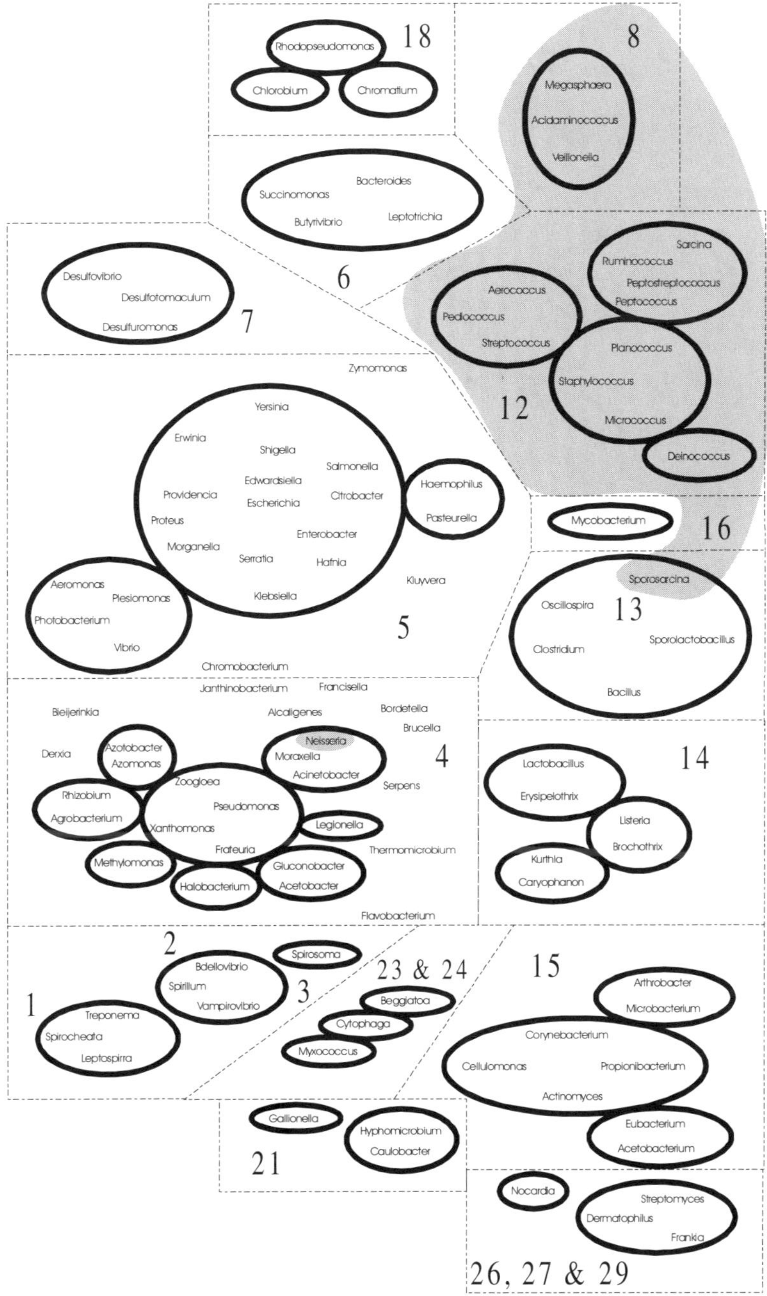

Figure 10.10 Atlas displaying as the shaded zones those bacterial genera that are commonly coccoid-shaped.

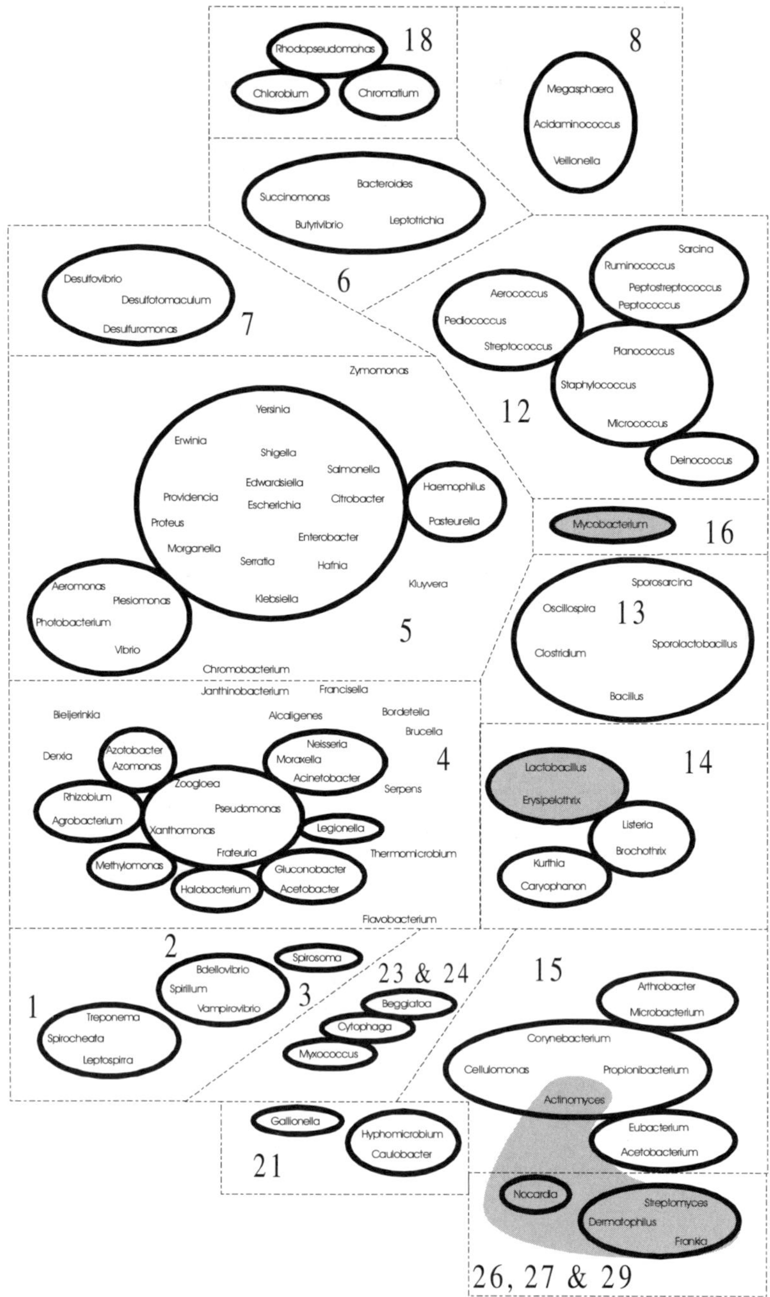

Figure 10.11 Atlas displaying as the shaded zones those bacterial genera that commonly are extensively filamentous or mycelial.

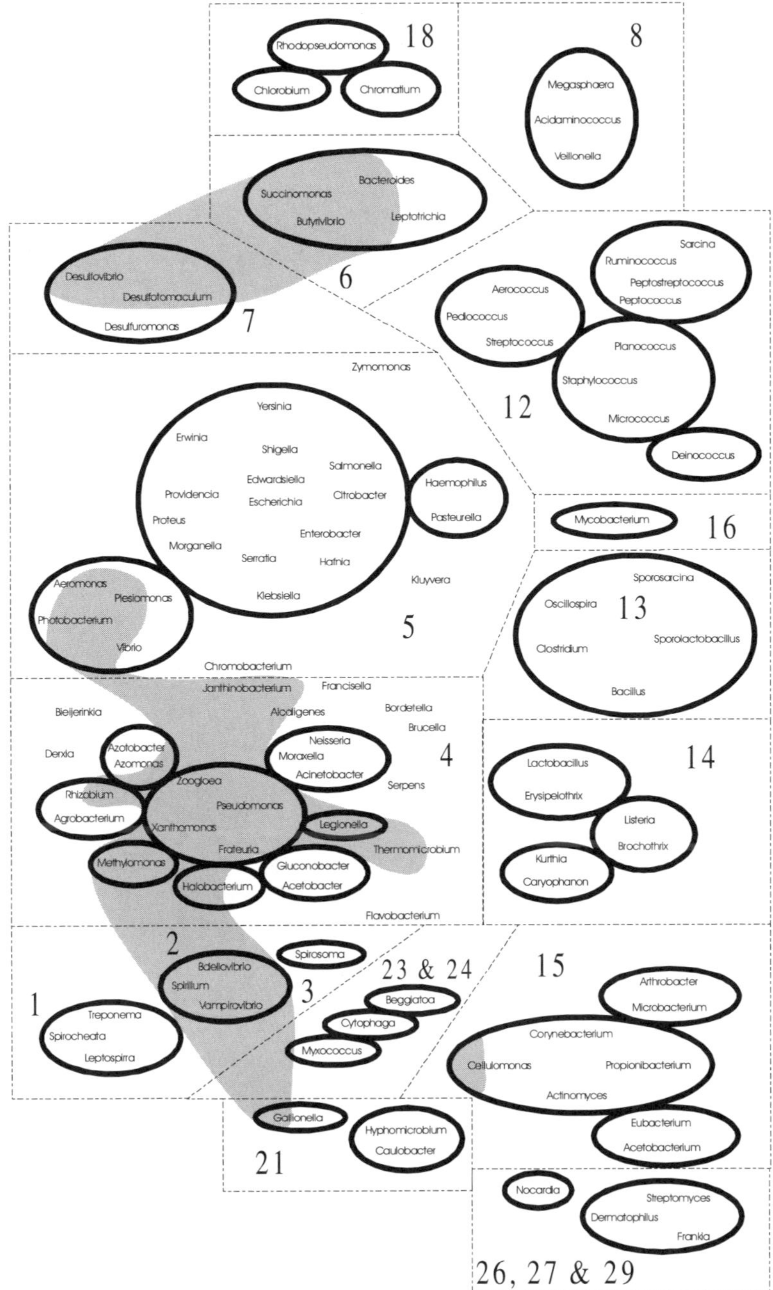

Figure 10.12 Atlas displaying as the shaded zones those bacterial genera that commonly have some form of polar flagellation.

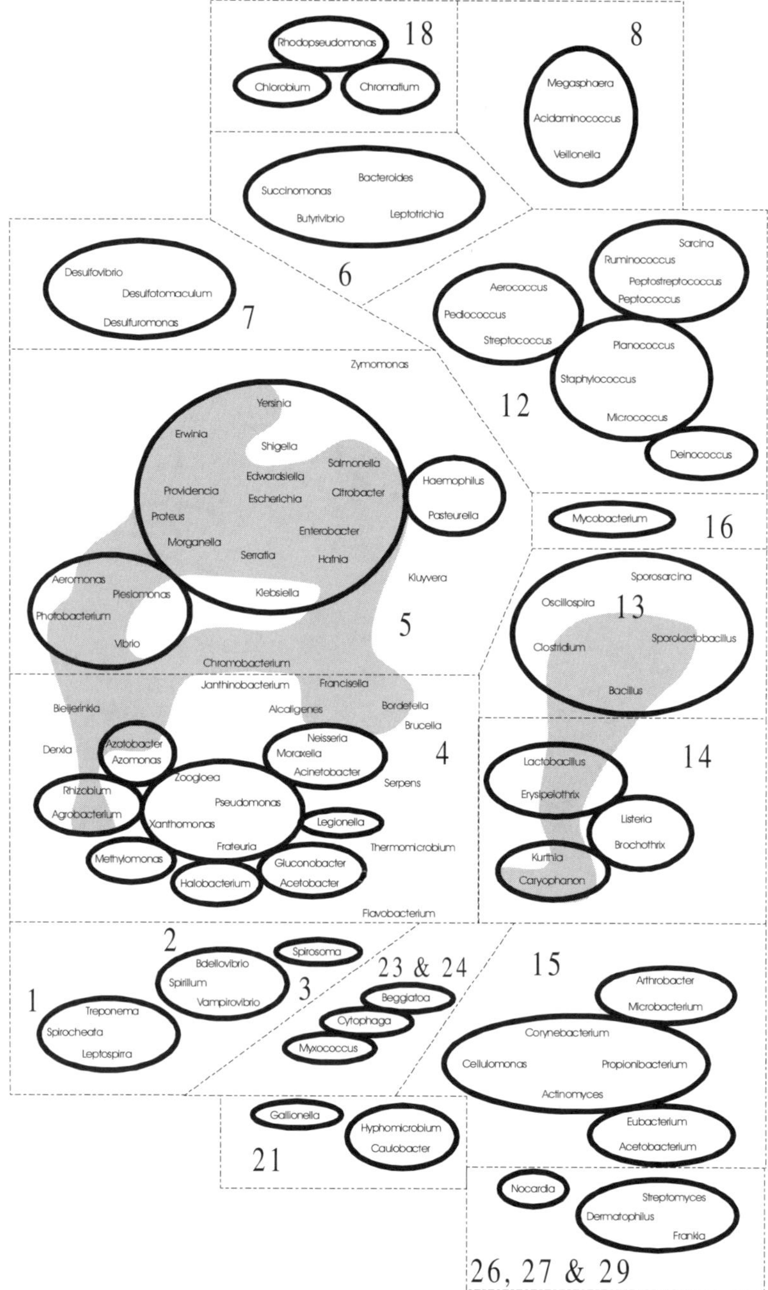

Figure 10.13 Atlas displaying as the shaded zones those bacterial genera that possess some form of peritrichous flagella pattern.

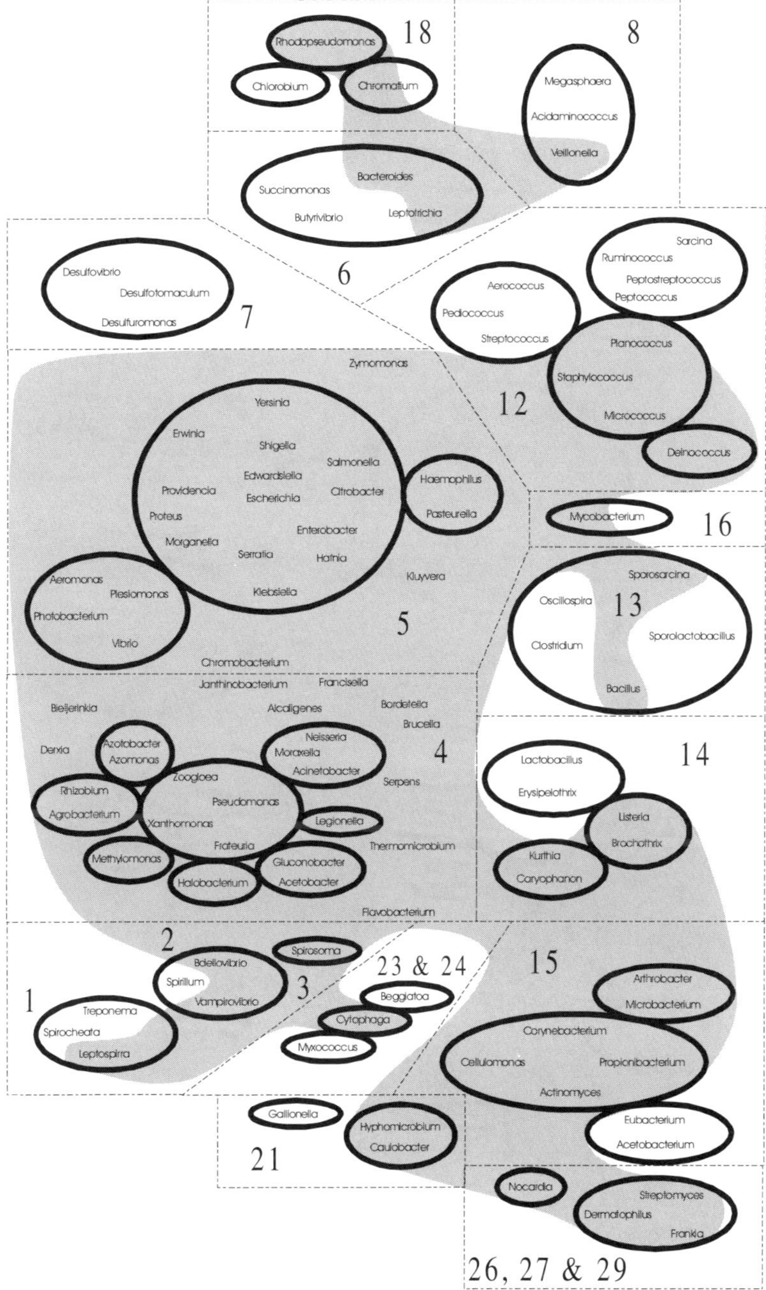

Figure 10.14 Atlas displaying as the shaded zones those bacterial genera that are commonly catalase positive.

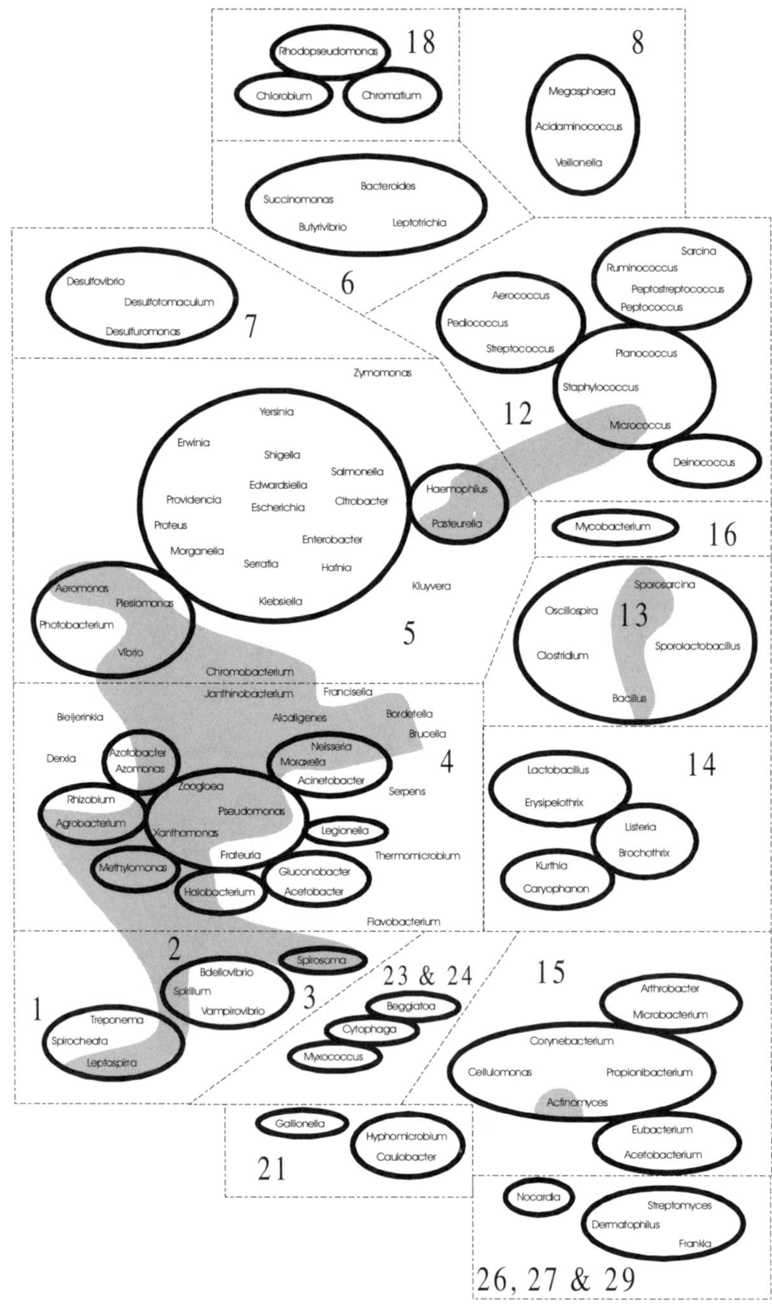

Figure 10.15 Atlas displaying as the shaded zones those bacterial genera that are oxidase positive.

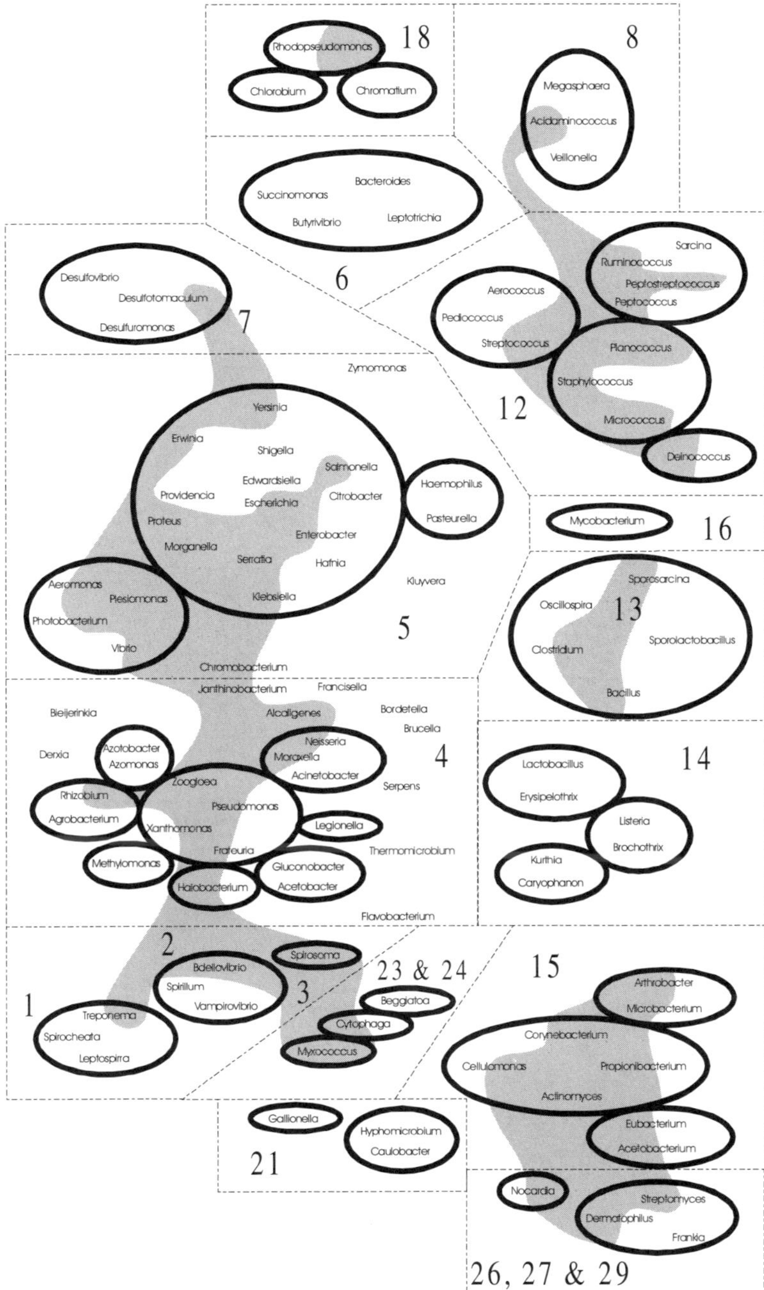

Figure 10.16 Atlas displaying as the shaded zones those bacterial genera that are commonly proteolytic.

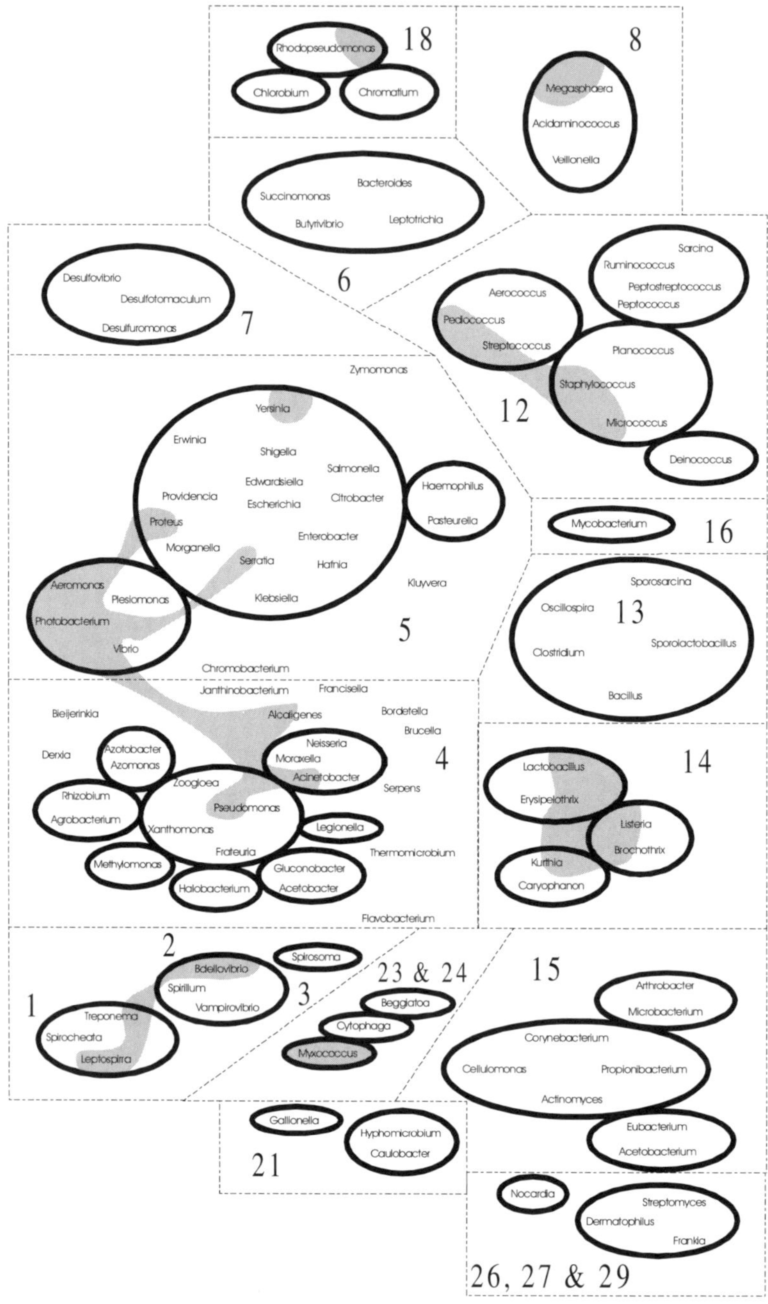

Figure 10.17 Atlas displaying as the shaded zones those bacterial genera that are commonly lipolytic.

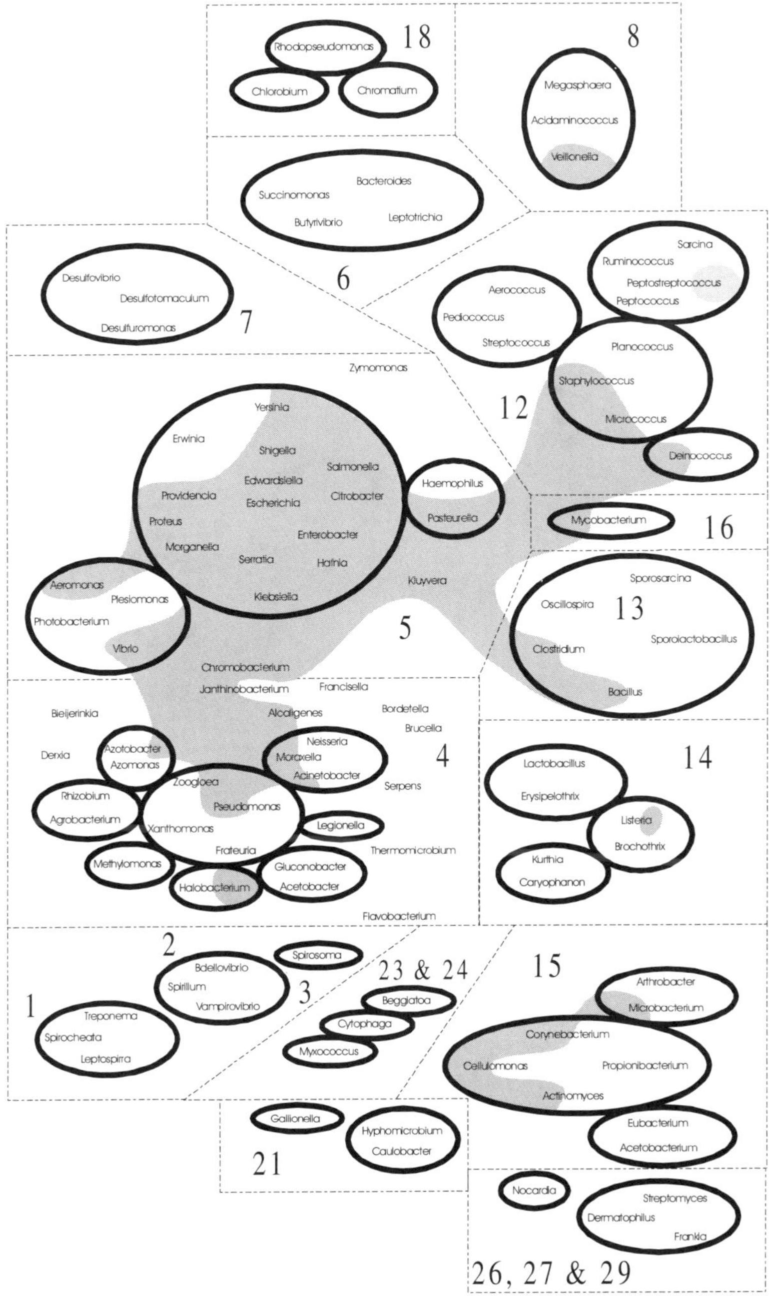

Figure 10.18 Atlas displaying as the shaded zones those bacterial genera that are able to respire or otherwise reduce nitrates (denitrification).

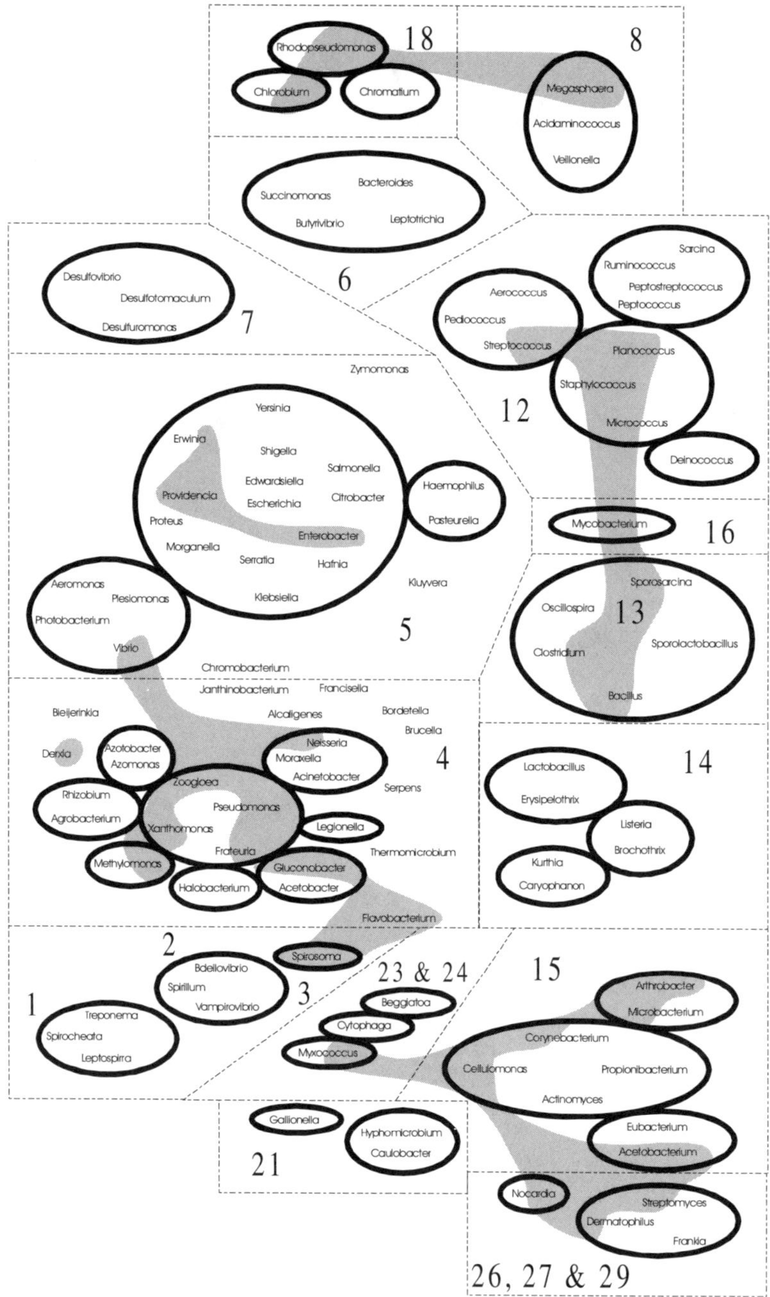

Figure 10.19 Atlas displaying as the shaded zones those bacterial genera that frequently have a yellow pigmentation particularly later on in the maturation of growth.

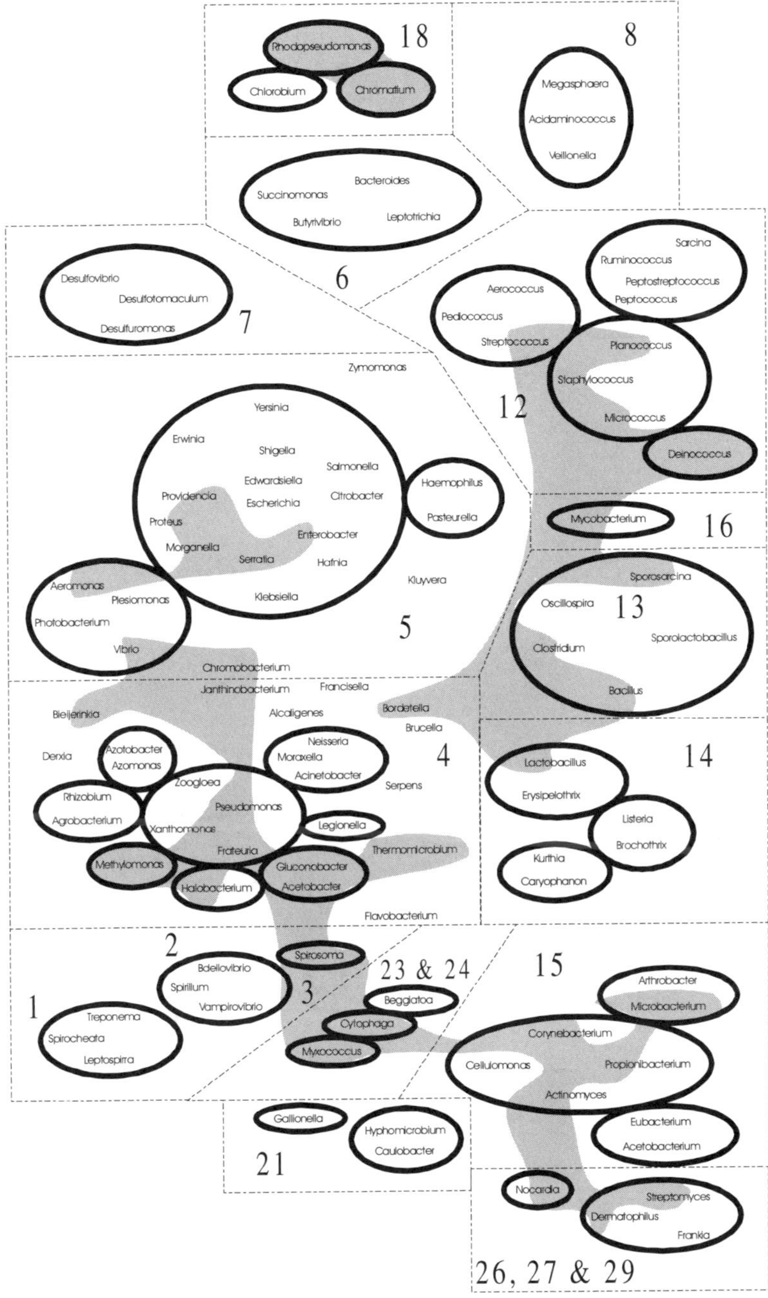

Figure 10.20 Atlas displaying as the shaded zones those bacterial genera that frequently have a red, orange or pink pigmentation particularly later on in the maturation of growth.

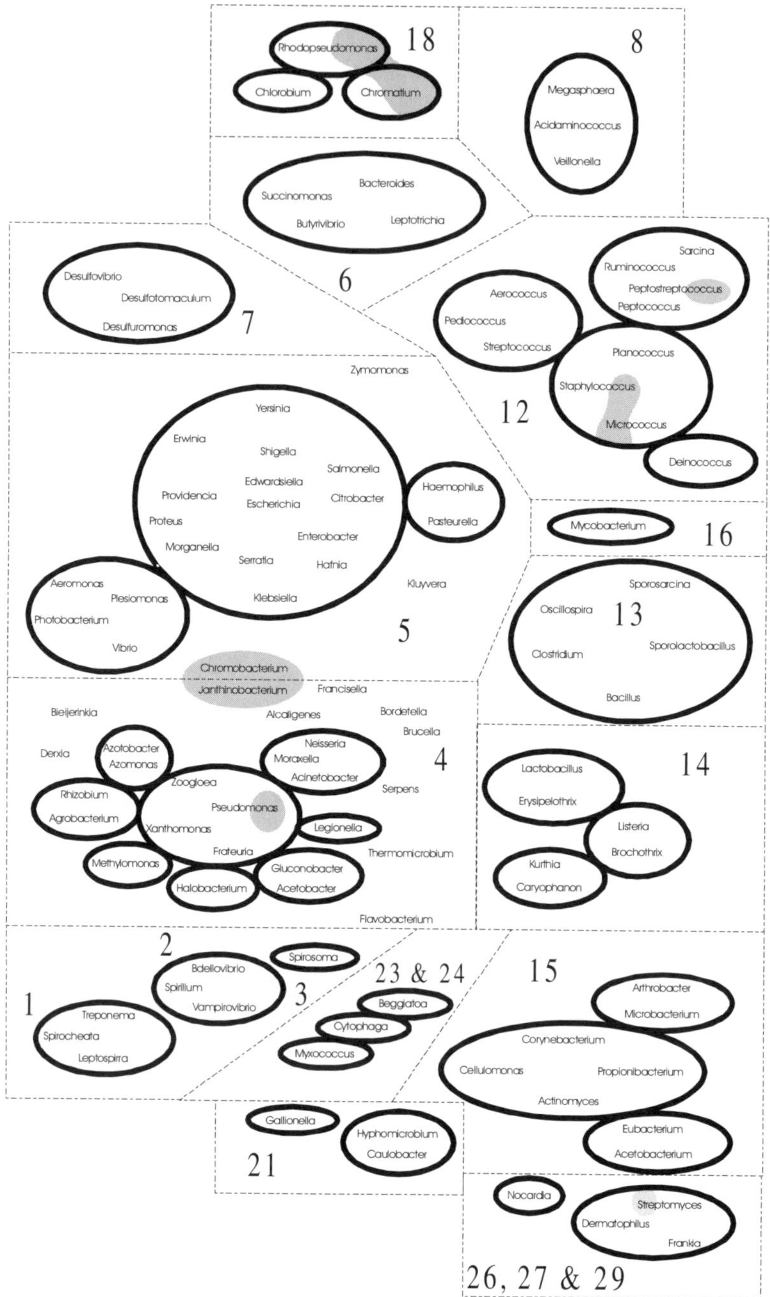

Figure 10.21 Atlas displaying as the shaded zones those bacterial genera that frequently have a purple or violet pigmentation particularly later on in the maturation of growth.

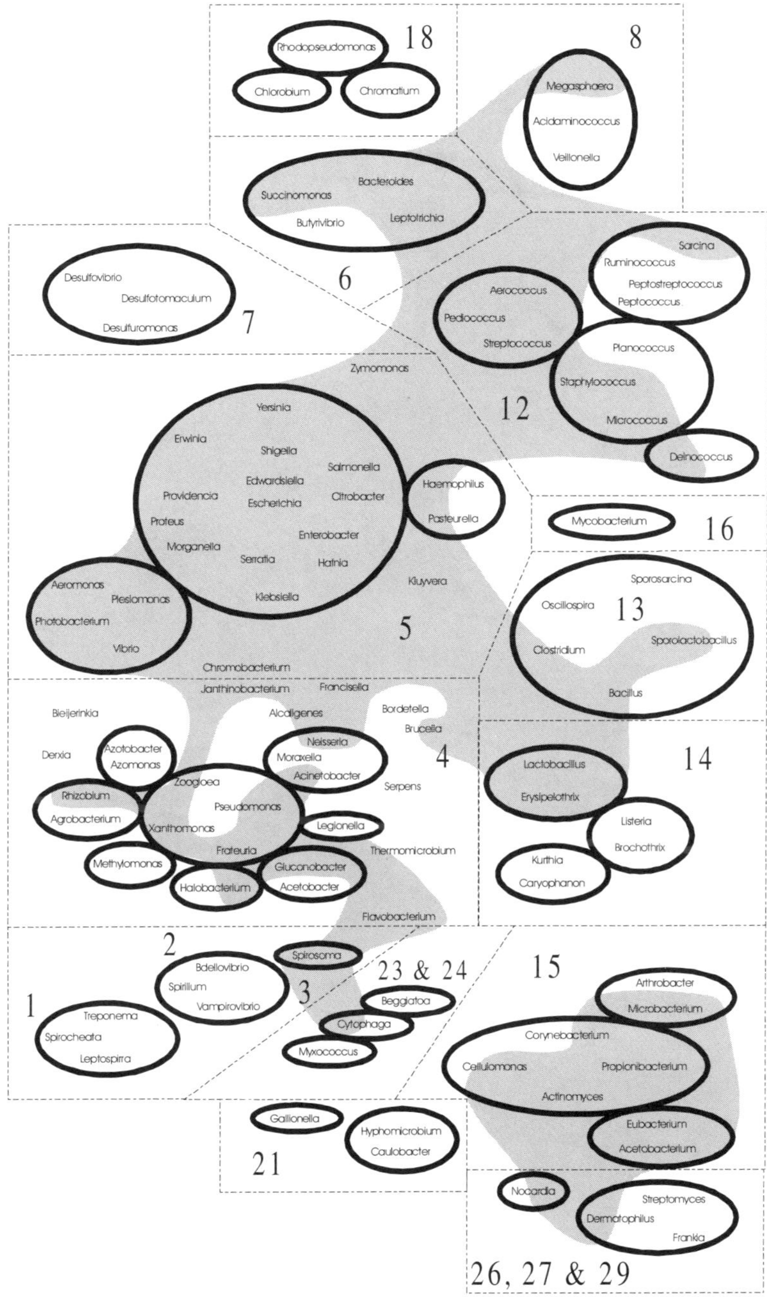

Figure 10.22 Atlas displaying as the shaded zones those bacterial genera that frequently degrade glucose with acidic end products.

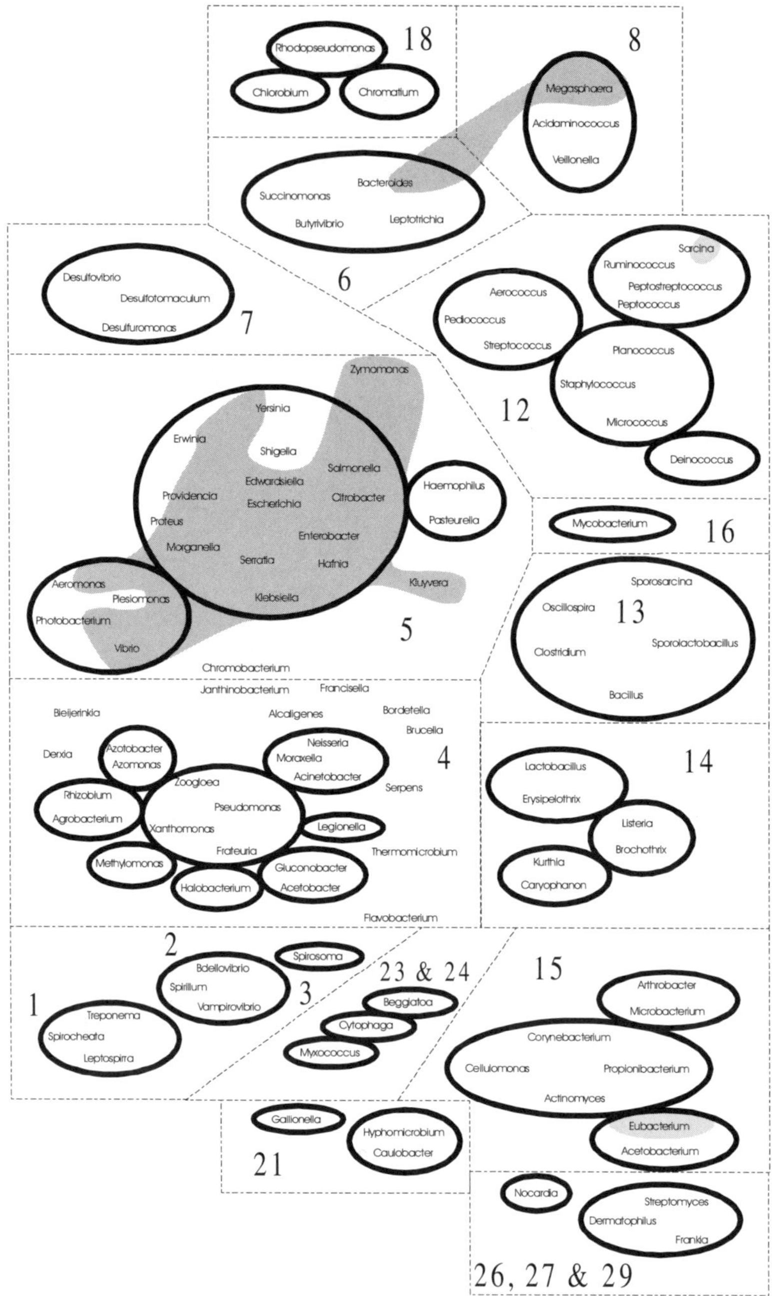

Figure 10.23 Atlas displaying as the shaded zones those bacterial genera that frequently degrade glucose with gaseous end products (dominantly carbon dioxide and hydrogen).

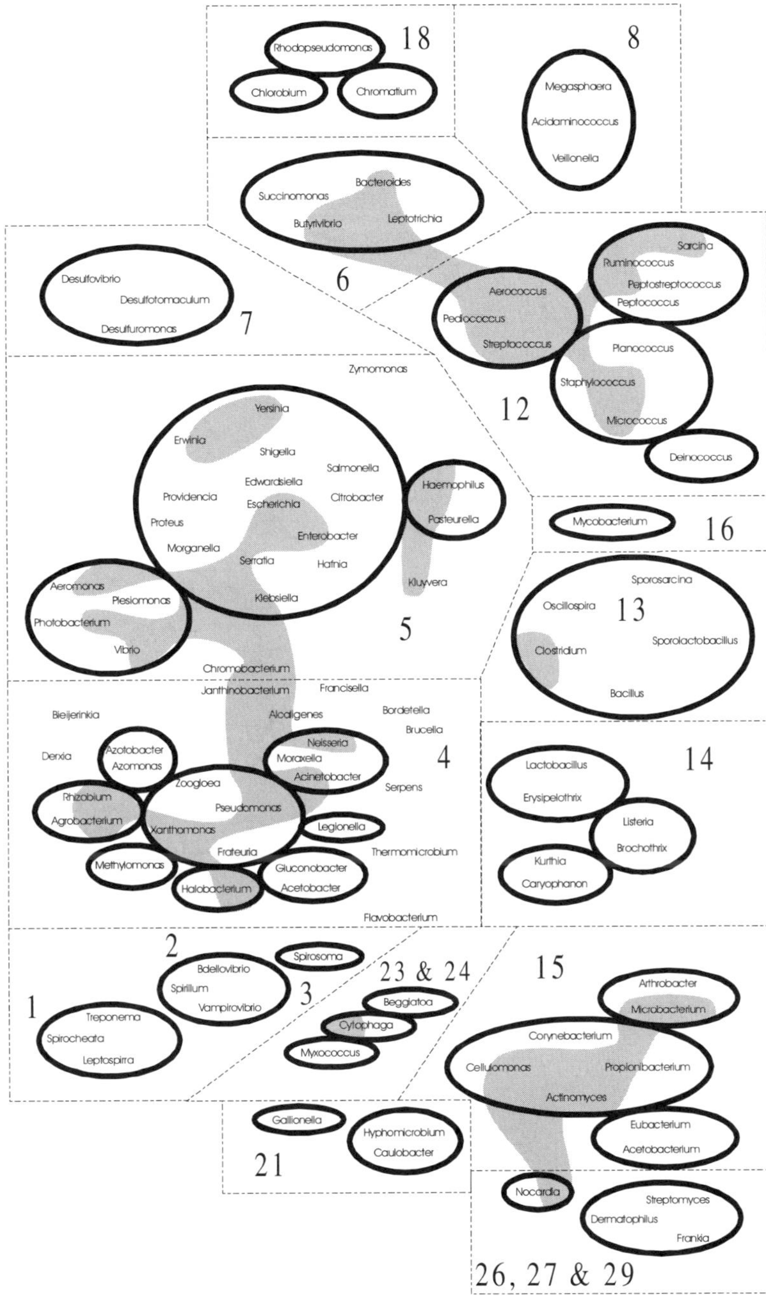

Figure 10.24 Atlas displaying as the shaded zones those bacterial genera that frequently degrade lactose with the production of acidic end products.

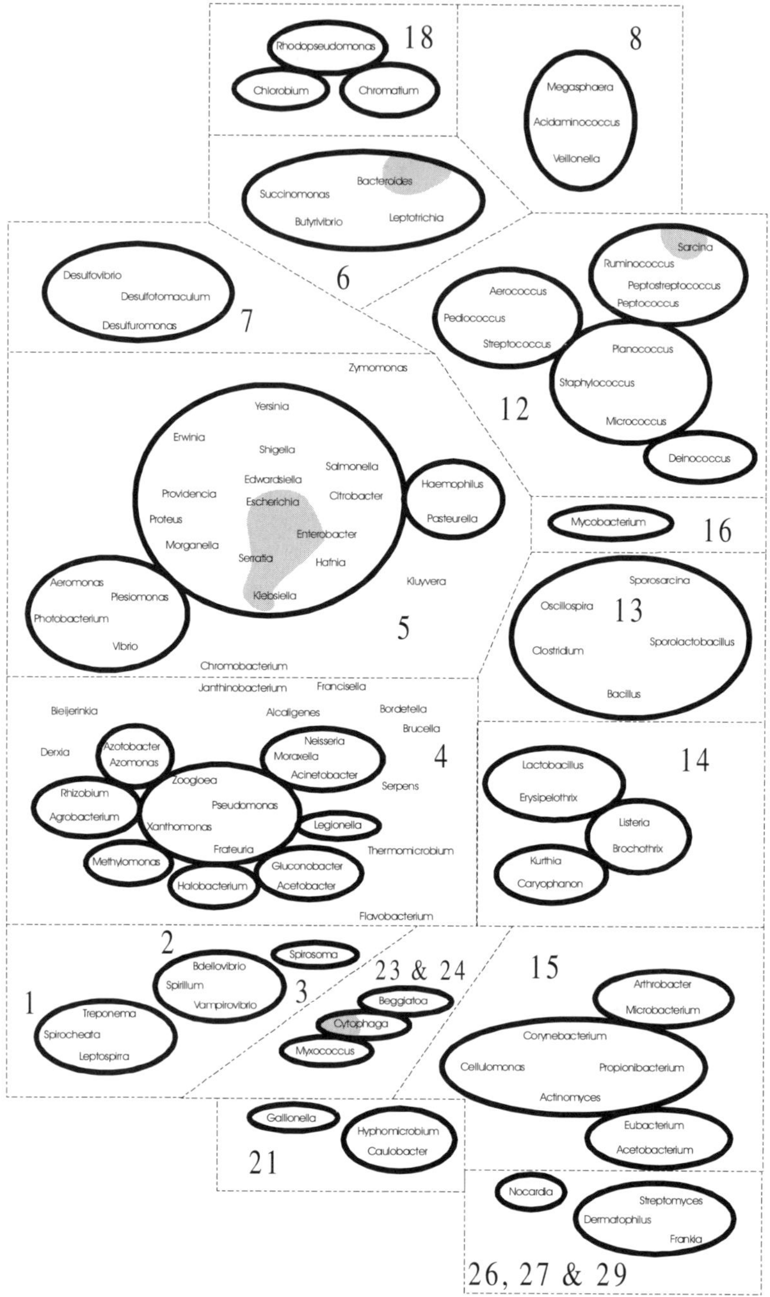

Figure 10.25 Atlas displaying as the shaded zones those bacterial genera that frequently degrade lactose with gaseous end products (dominantly carbon dioxide and hydrogen).

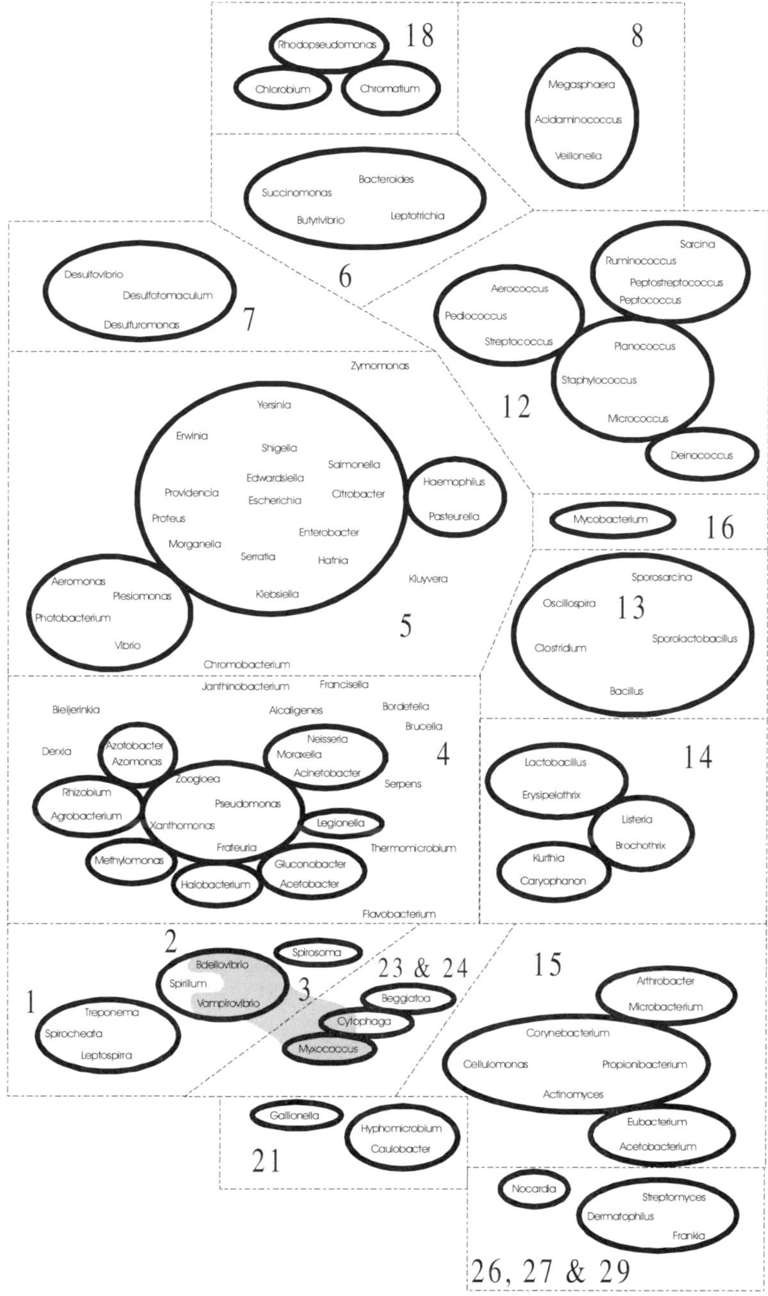

Figure 10.26 Atlas displaying as the shaded zones those bacterial genera that are frequently parasitic on other microorganisms.

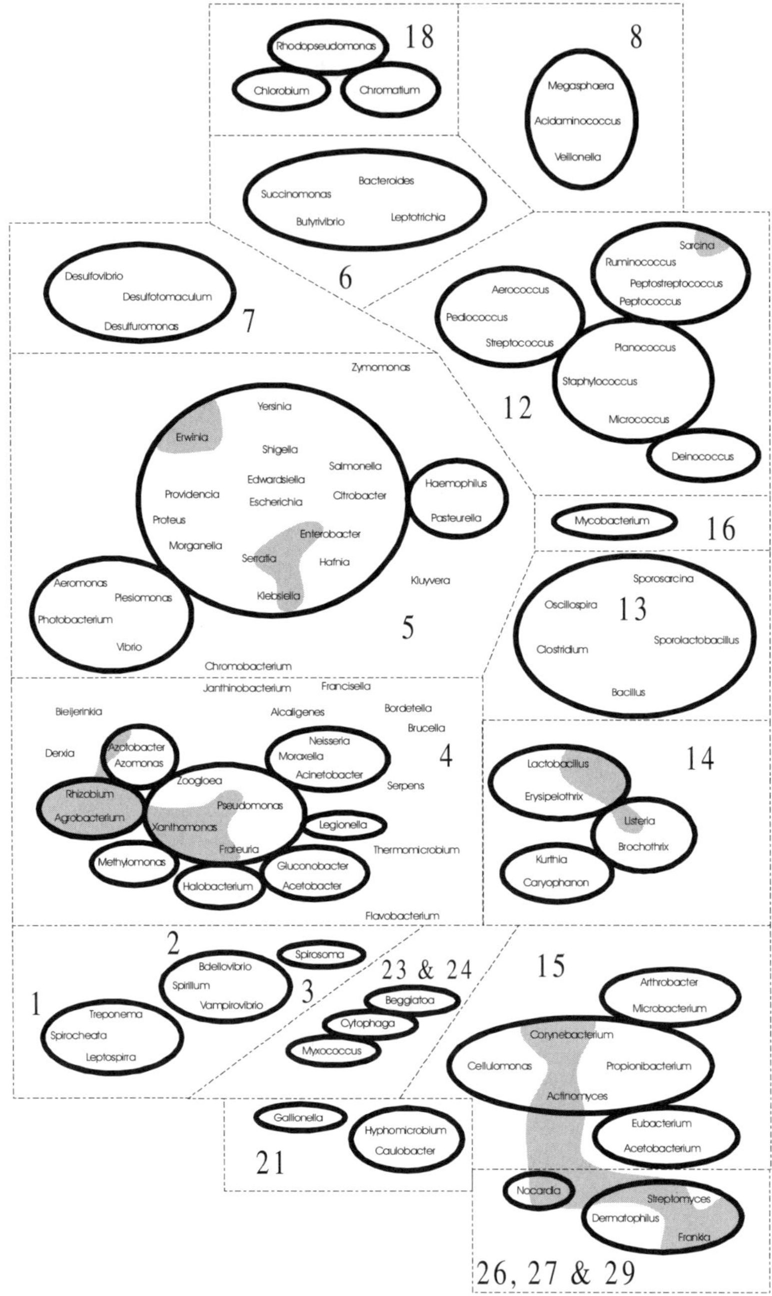

Figure 10.27 Atlas displaying as the shaded zones those bacterial genera that are frequently parasitic on plants.

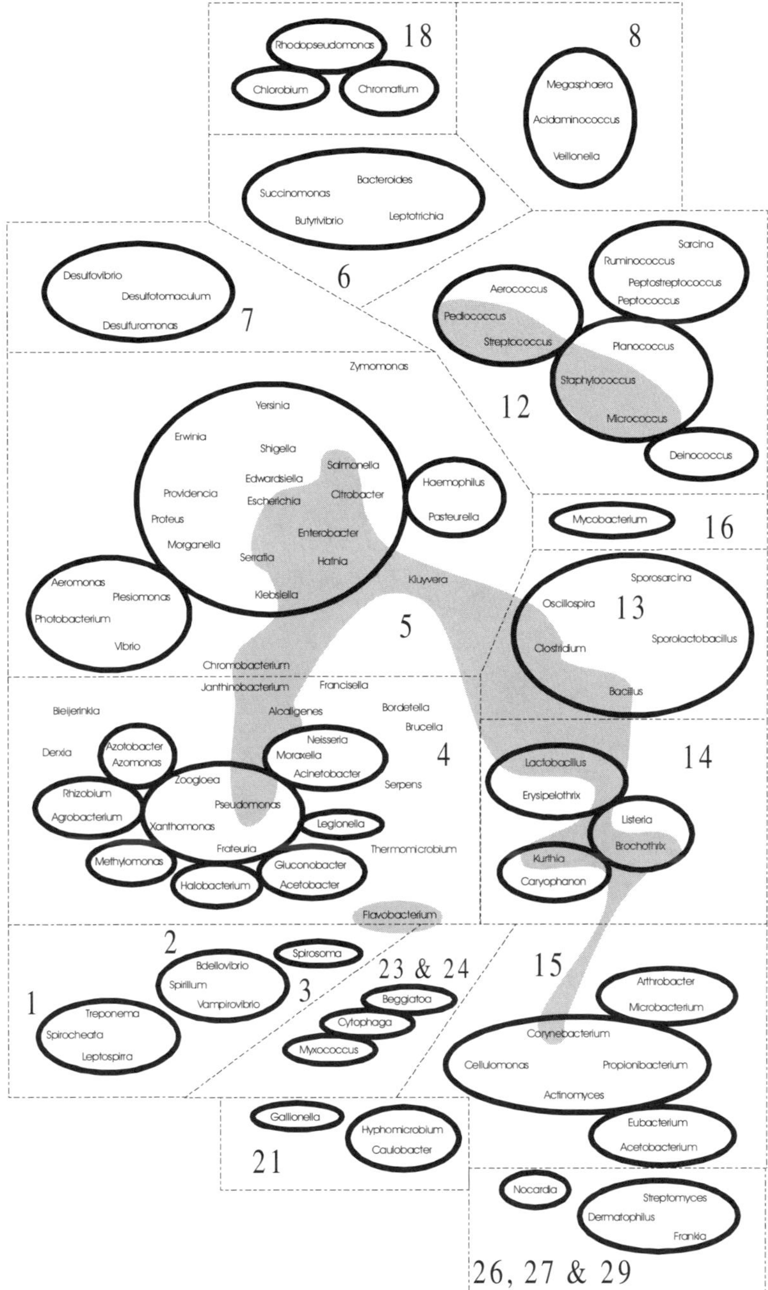

Figure 10.28 Atlas displaying as the shaded zones those bacterial genera that include species known to cause spoilage in milk and dairy products.

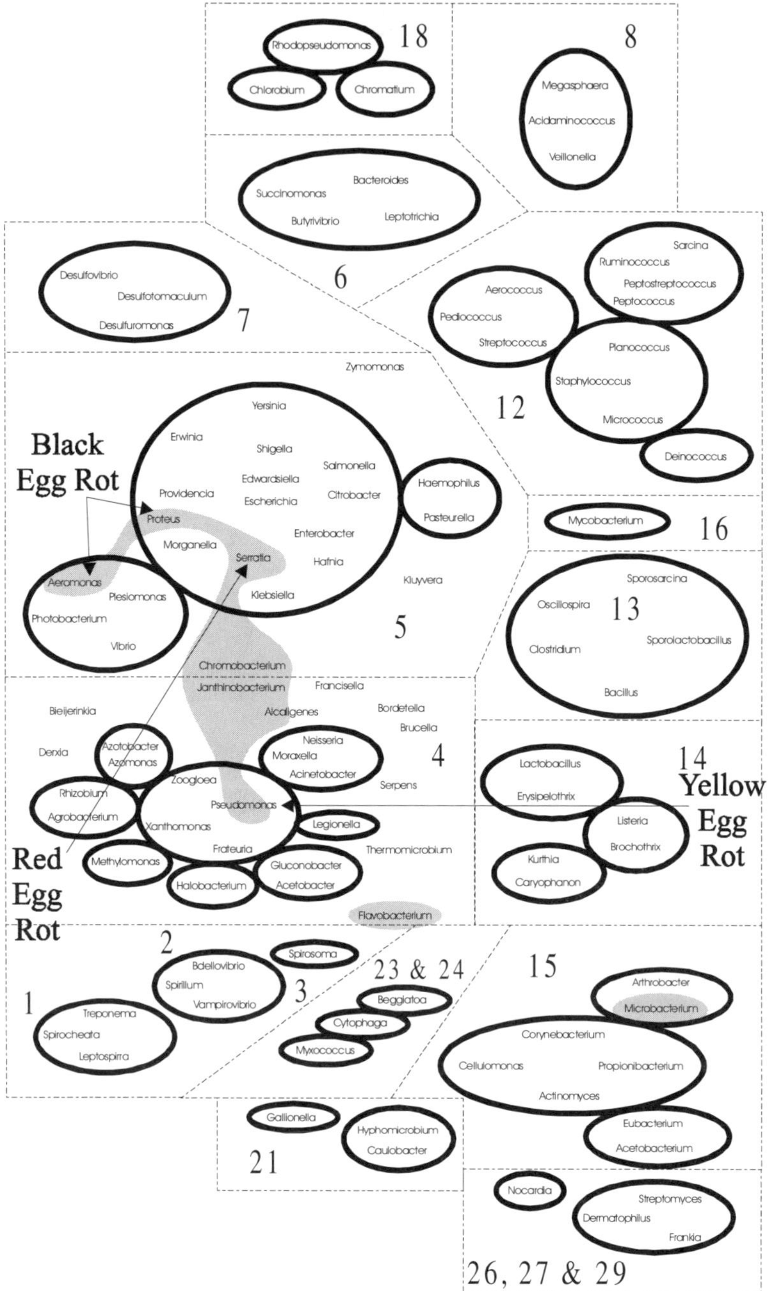

Figure 10.29 Atlas displaying by partial box and arrow those bacterial genera that include species known to cause spoilage in egg and poultry products.

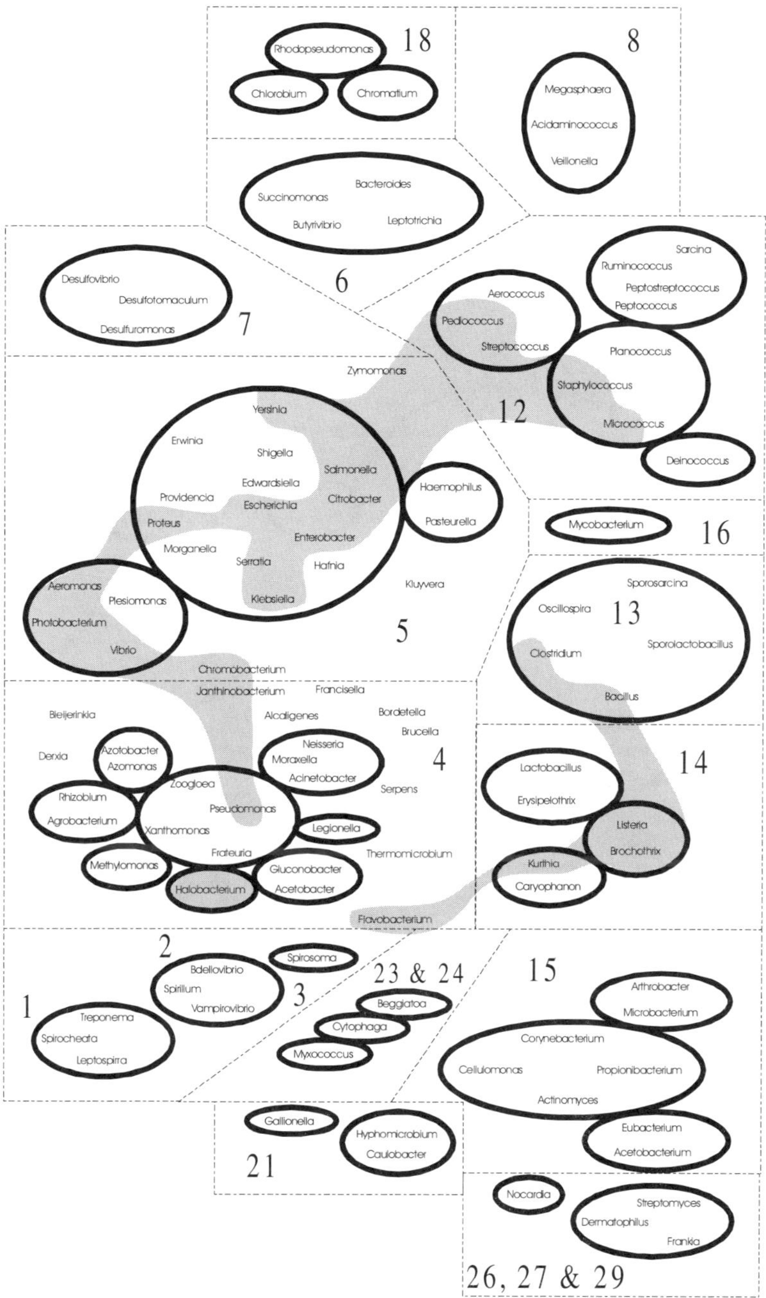

Figure 10.30 Atlas displaying by partial box and arrow those bacterial genera that include species known to cause spoilage in meat products.

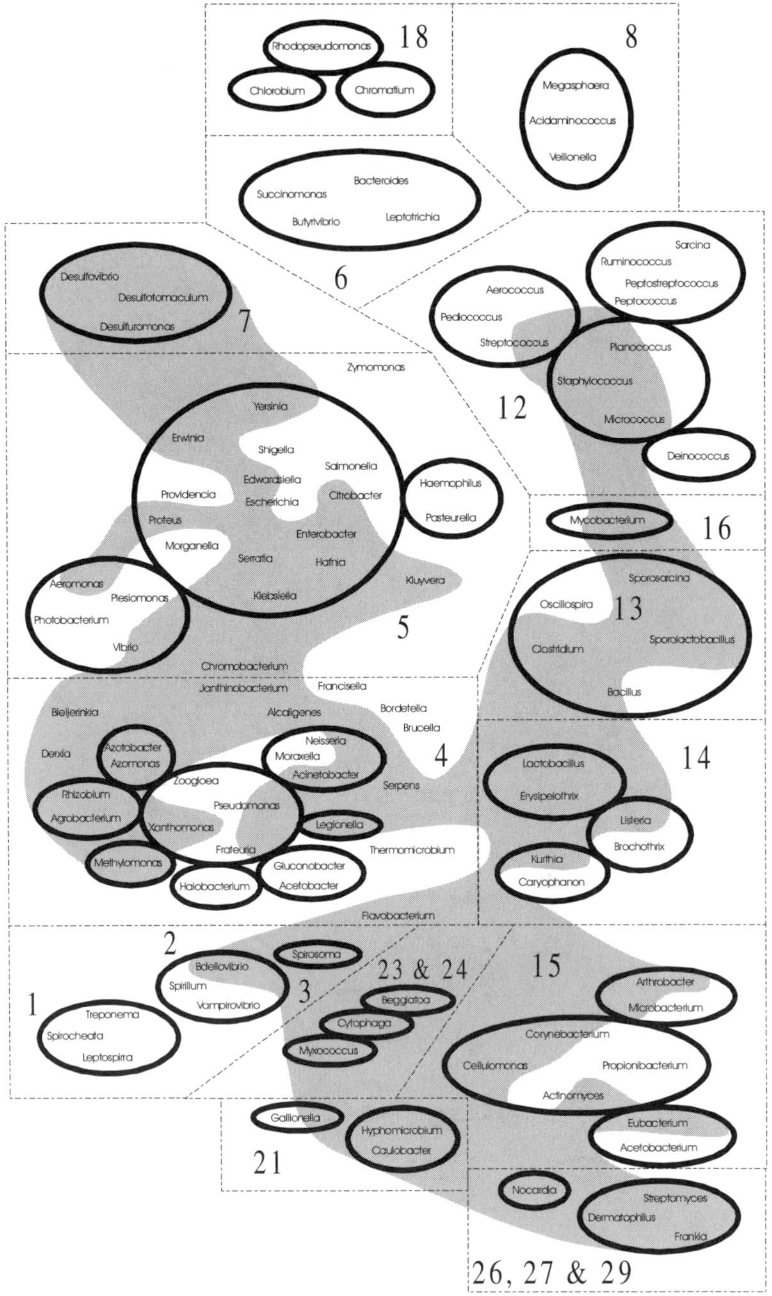

Figure 10.31 Atlas displaying as the shaded zones those bacterial genera that include species known to commonly inhabit soils.

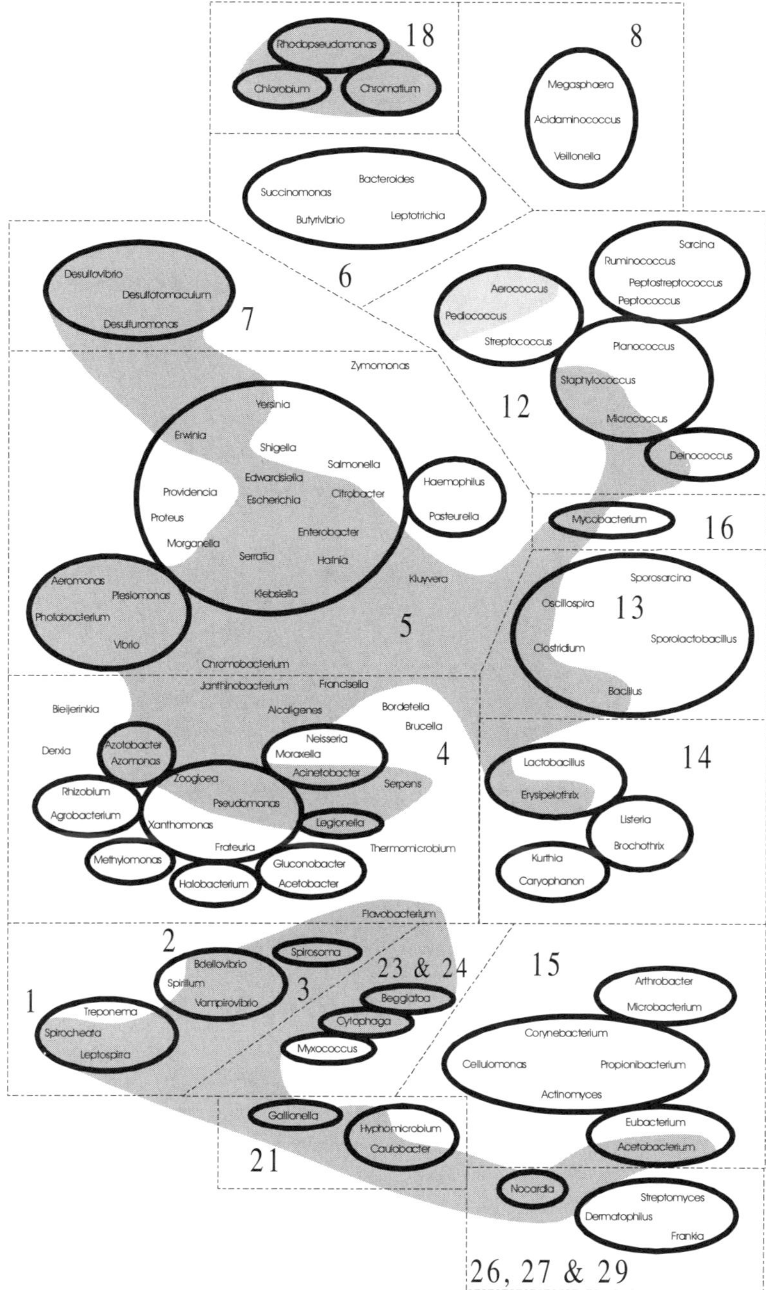

Figure 10.32 Atlas displaying as the shaded zones those bacterial genera that include species known to inhabit water.

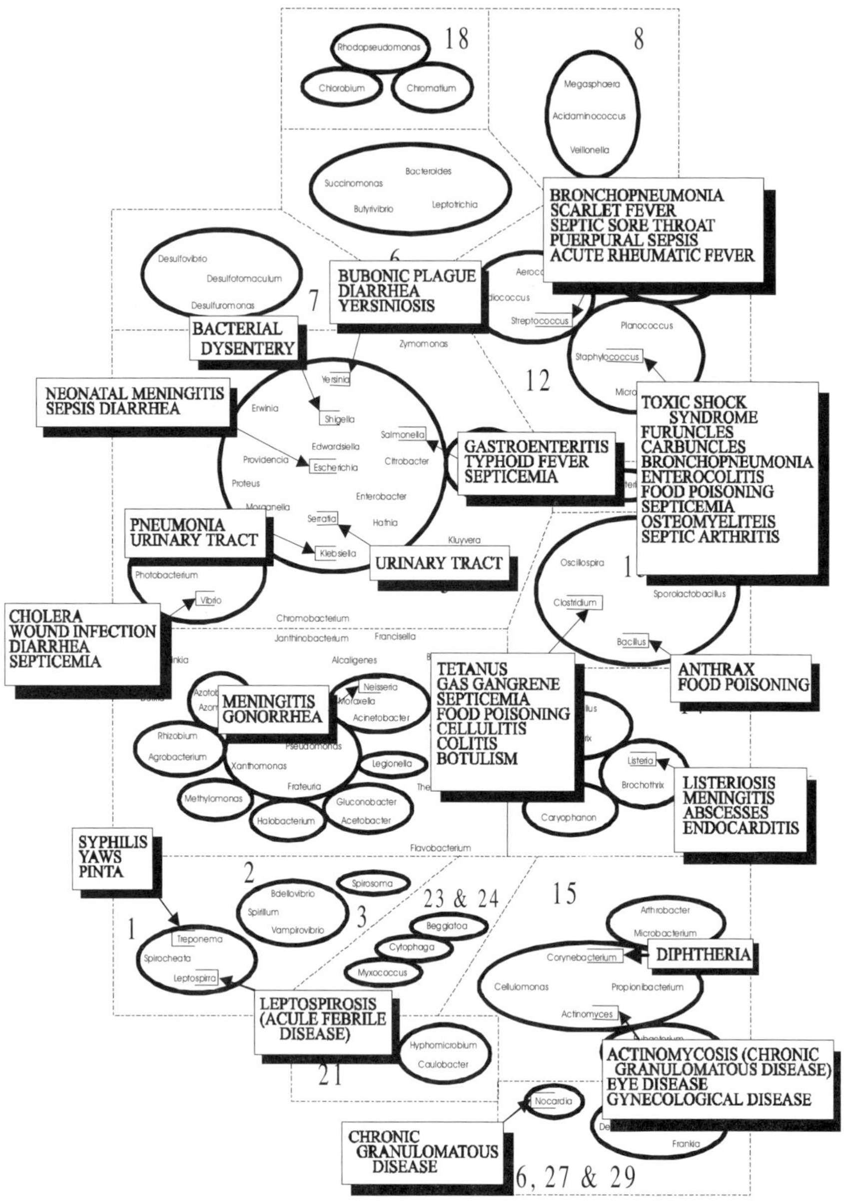

Figure 10.33 Atlas displaying as the shaded zones those bacterial genera that include some of the species known to cause infectious diseases in the human species (see also Figures 10.34 and 10.35).

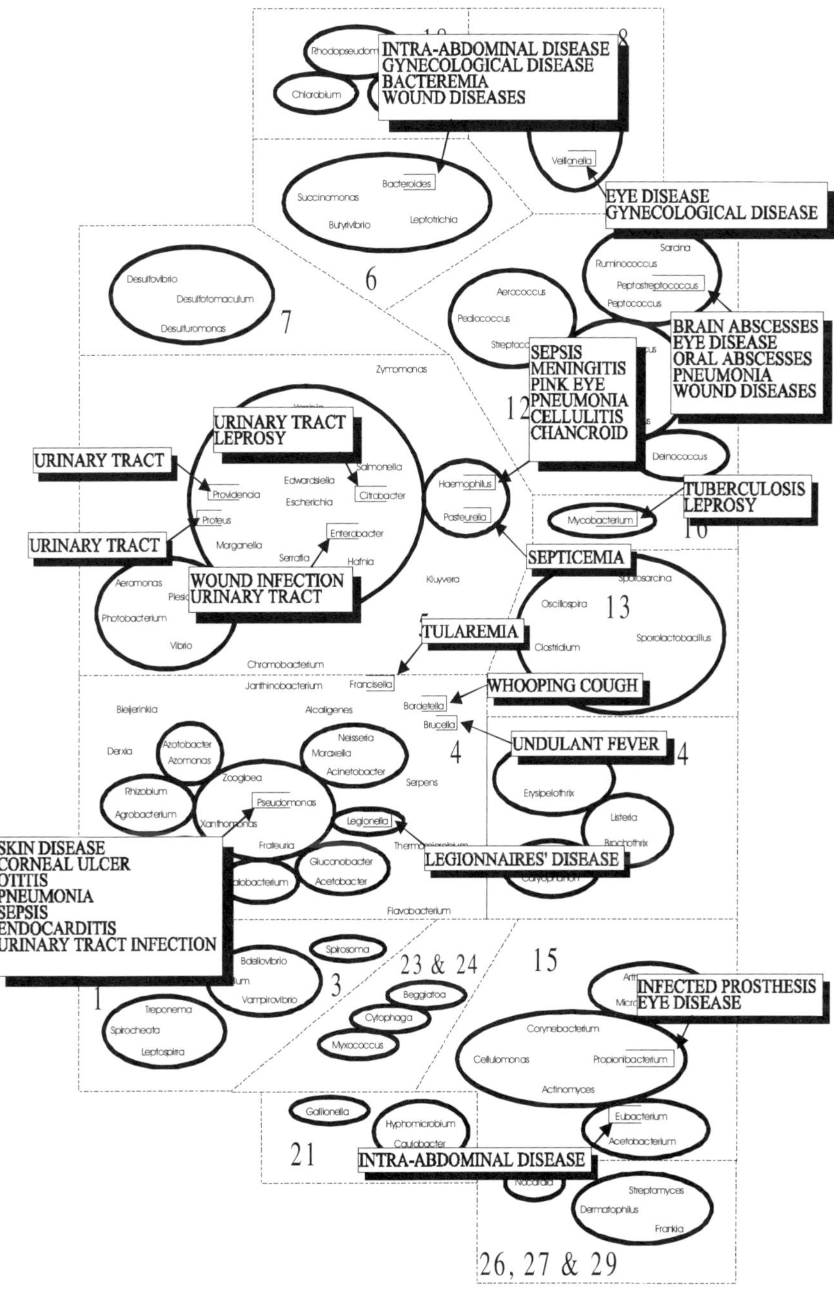

Figure 10.34 Atlas displaying as the shaded zones those bacterial genera that include some of the species known to cause infectious diseases in the human species (see also Figures 10.33 and 10.35).

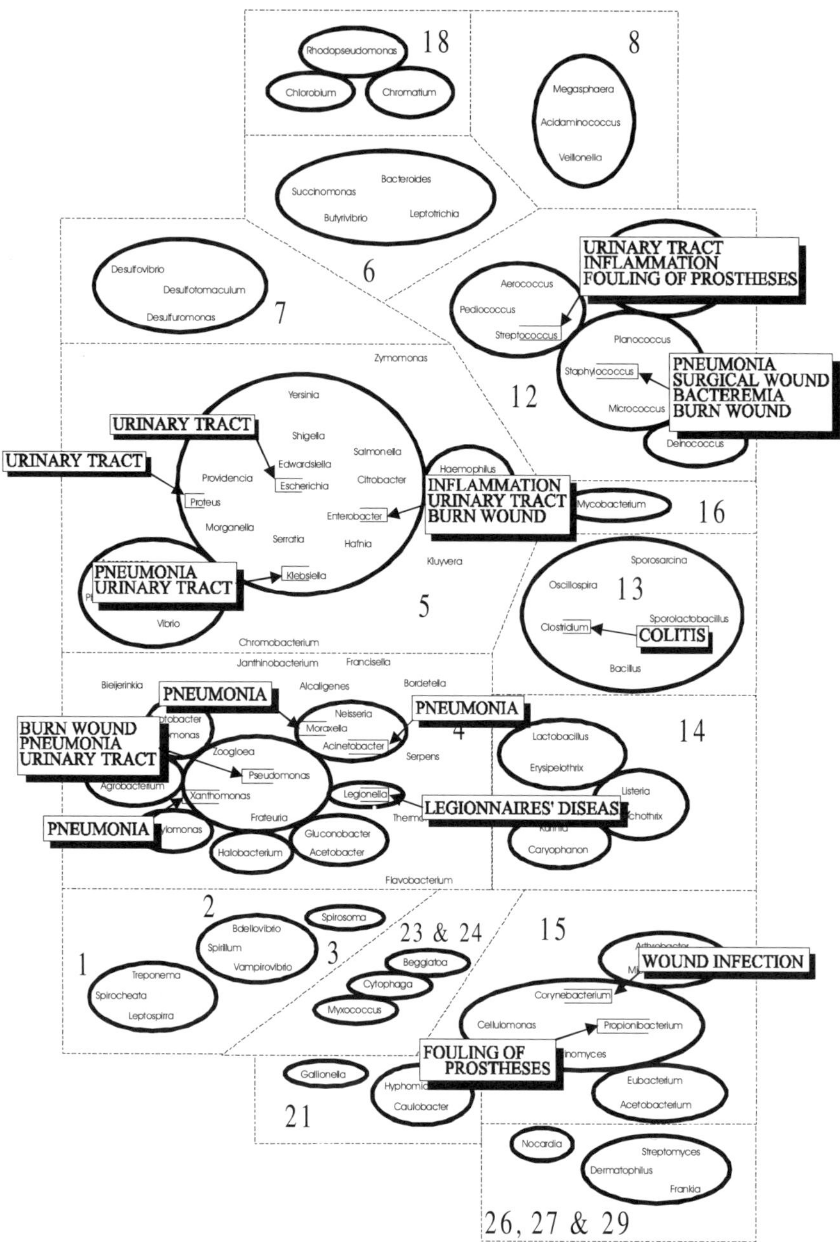

Figure 10.35 Atlas displaying by partial box and arrow those bacterial genera that include some of the species known to cause hospital-induced nosocomial diseases in the human species (see also Figures 10.33 and 10.34 for the species that cause infectious diseases).

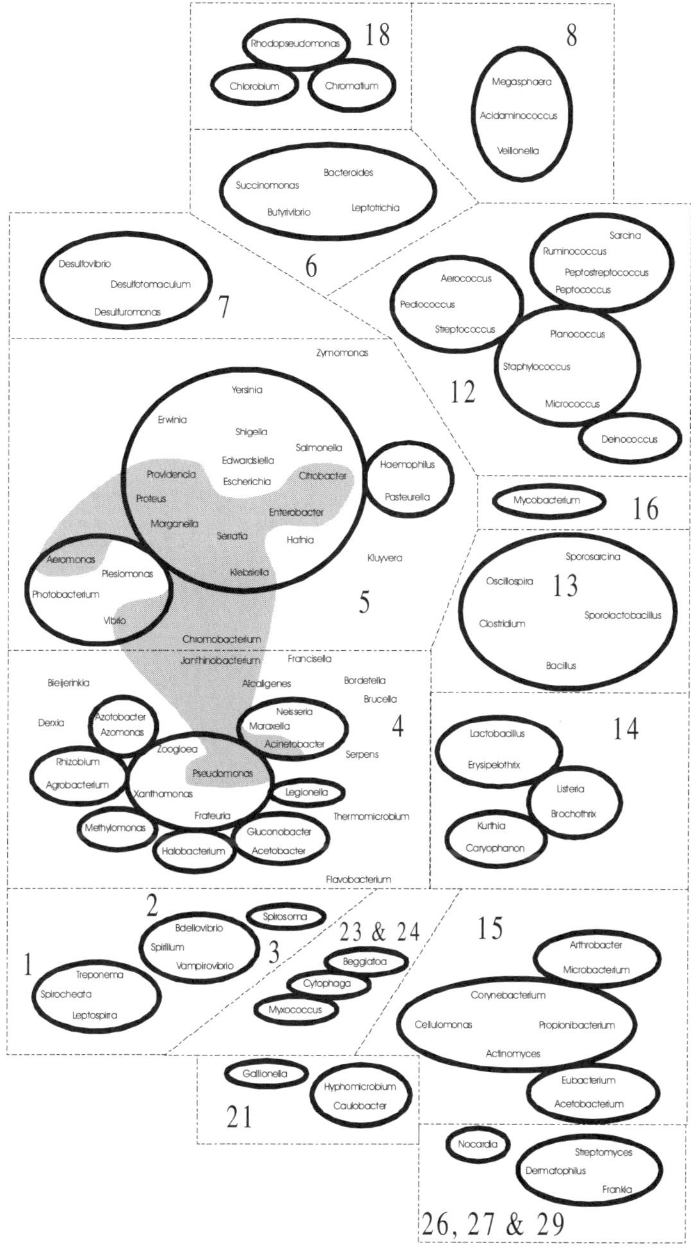

Figure 10.36 Common bacterial consortial complexes in the iron-related bacteria (IRB) BART that cause an RPS involving (in which sequence) the following reaction codes FO - GC - BL are likely to contain many of the genera shaded with Section 4 pseudomonad bacteria tending to dominate the consortium.

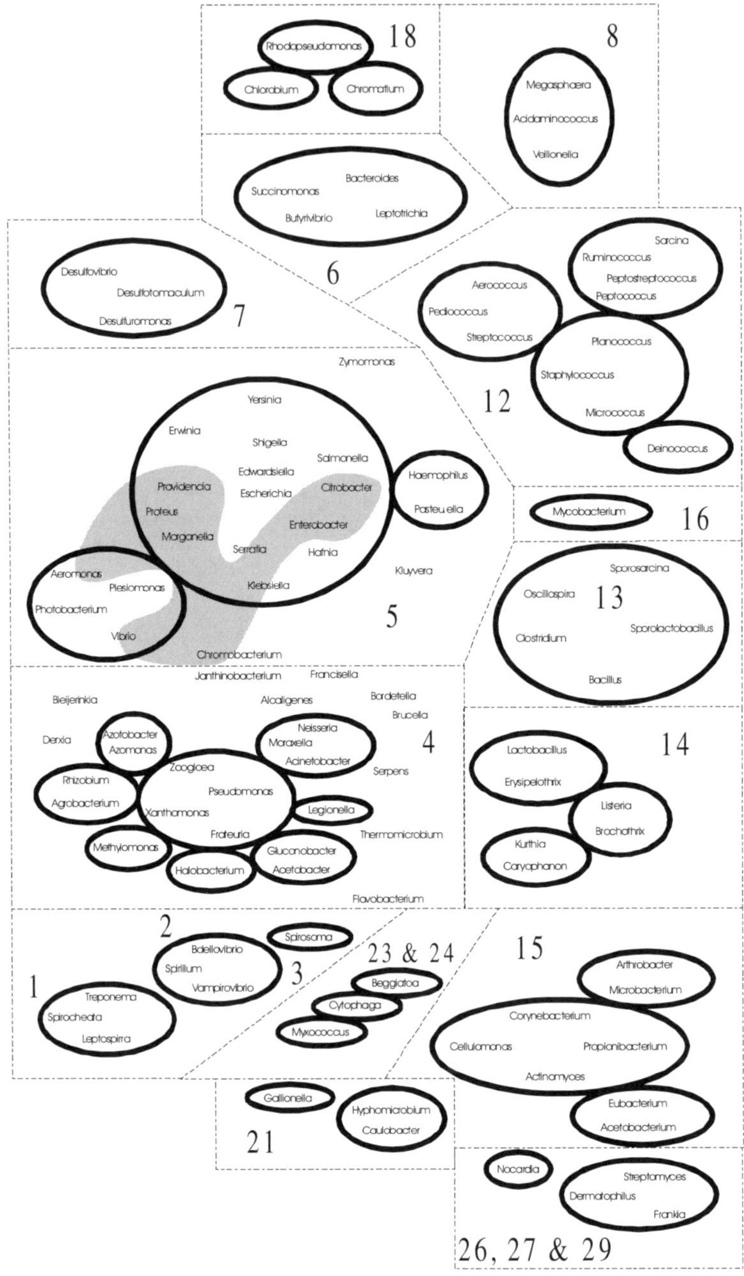

Figure 10.37 Common bacterial consortial complexes in the iron-related bacteria (IRB) BART cause an RPS involving (in which sequence) the following reaction codes FO - CL - RC are likely to contain many of the genera shaded with Section 5 enteric bacteria dominating the consortium.

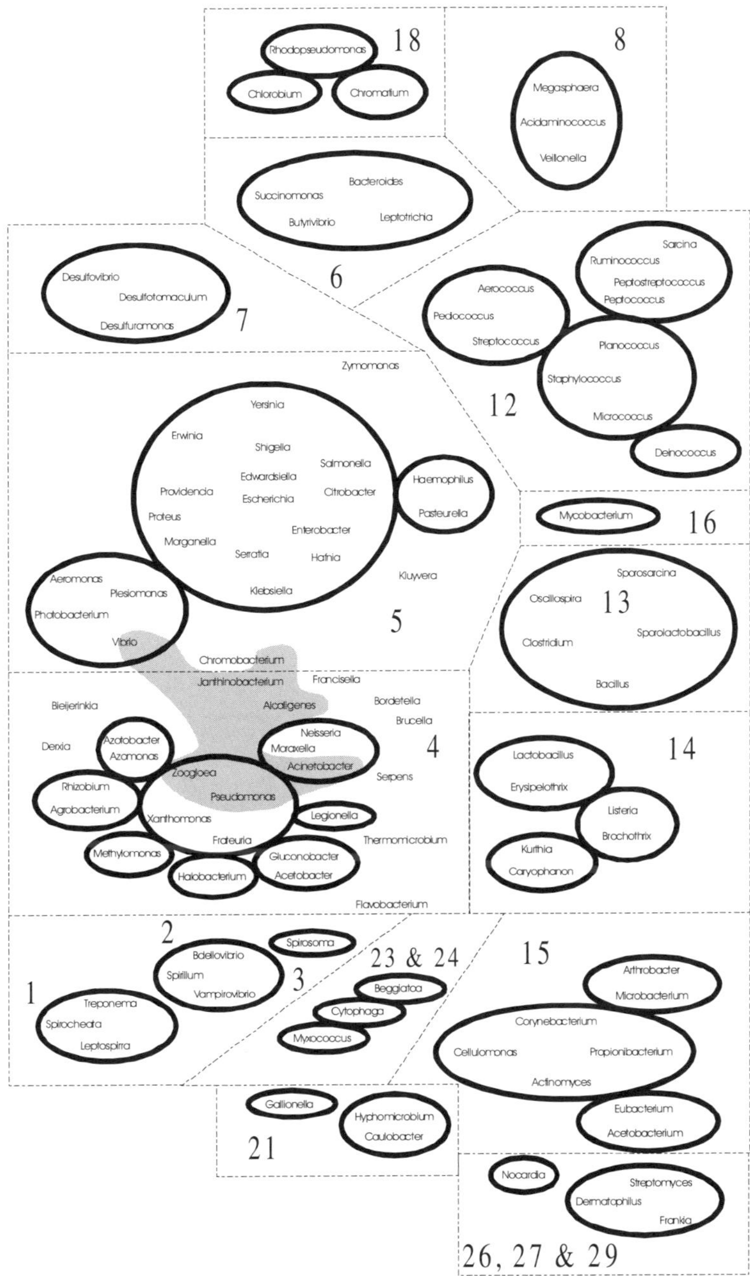

Figure 10.38 Common bacterial consortial complexes in the iron-related bacteria (IRB) BART cause an RPS involving (in which sequence) the following reaction codes GC possibly preceded by a CL are likely to contain many of the genera shaded with Section 4 pseudomonad bacteria dominant.

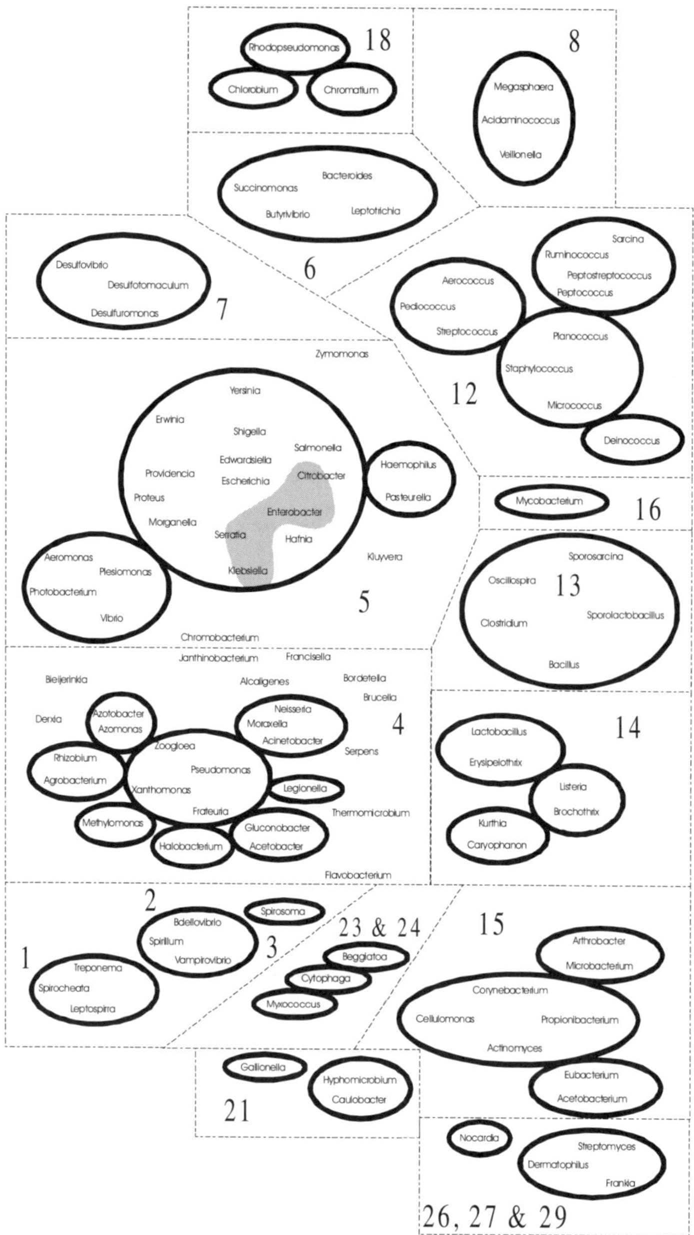

Figure 10.39 Common bacterial consortial complexes in the iron-related bacteria (IRB) BART cause an RPS involving (in which sequence) the following reaction codes CL - BG are likely to contain many of the genera shaded with Section 5 enteric bacteria with the genus *Enterobacter* tending to dominate the consortium.

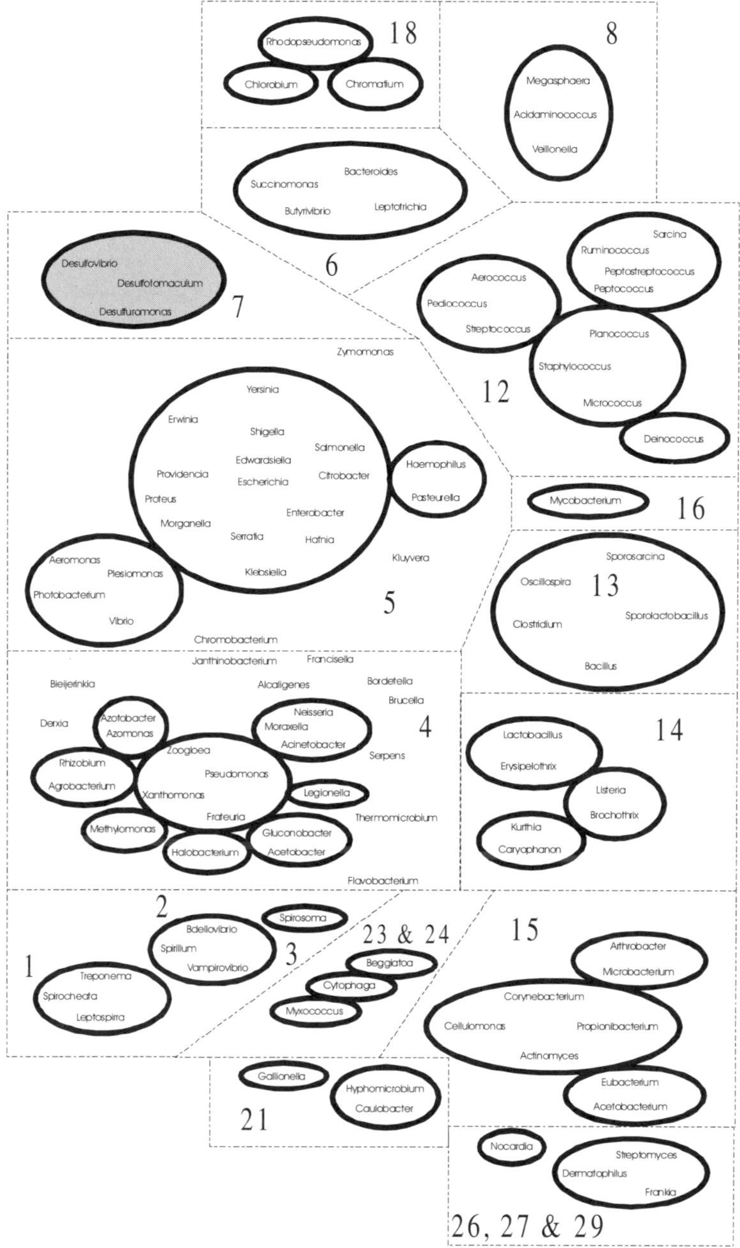

Figure 10.40 Common bacterial consortial (shaded) complexes in the sulfate-reducing bacteria (SRB) BART are dominated by the Section 7 sulfate-reducing bacteria and these commonly dominate the consortia. Note that *Desulforuromonas* will only be present when there is an elemental sulfur substrate available.

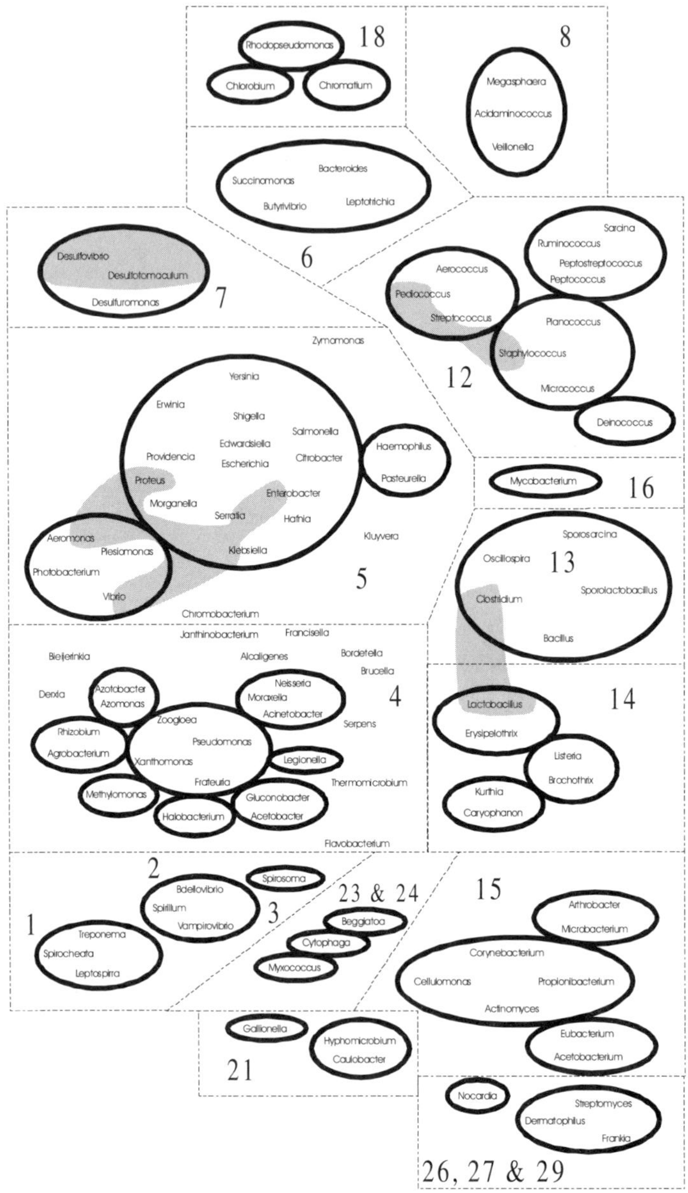

Figure 10.41 Common bacterial consortial (shaded) complexes in the sulfate-reducing bacteria (SRB) BART are dominated by the Section 7 sulfate-reducing bacteria and these commonly dominate the consortia. In the BB reaction where the blackening is in the base of the test vial, the likely genera to be present are shown in the atlas above.

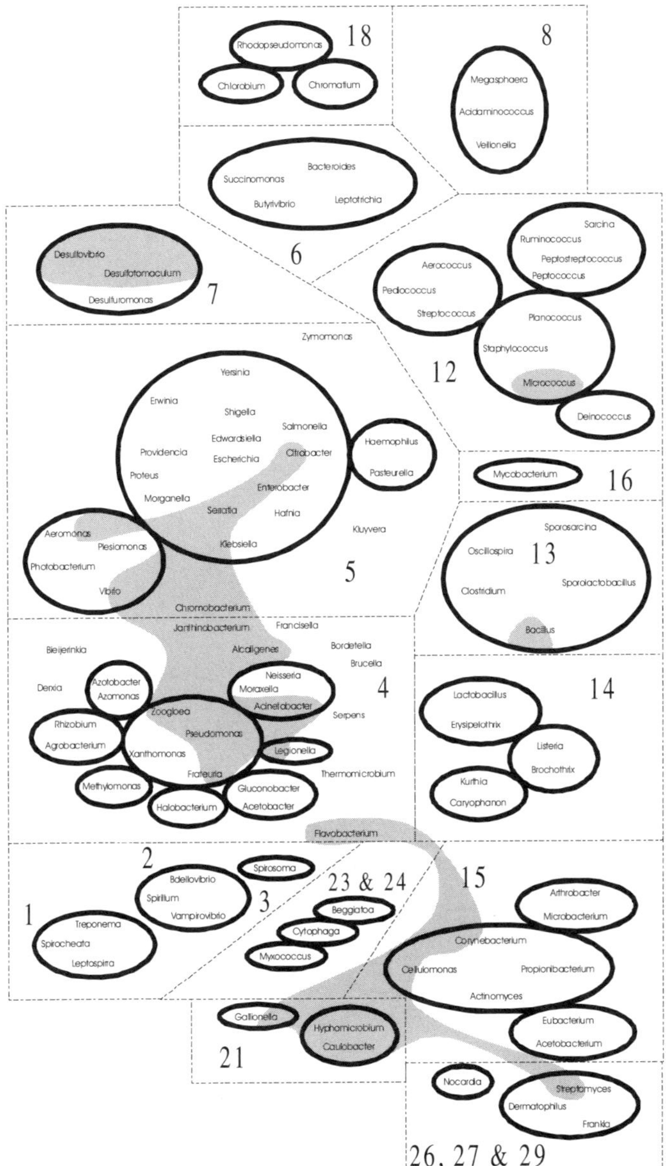

Figure 10.42 Common bacterial consortial (shaded) complexes in the sulfate-reducing bacteria (SRB) BART are dominated by the Section 7 sulfate-reducing bacteria and these commonly dominate the consortia. In the BT reaction where the blackening forms often in a granular manner around the ball in the test vial, the likely genera to be present are shown in the atlas above. Here, the SRB are involved in a primarily aerobic consortium and are surviving in the deeper layers of the biofilms.

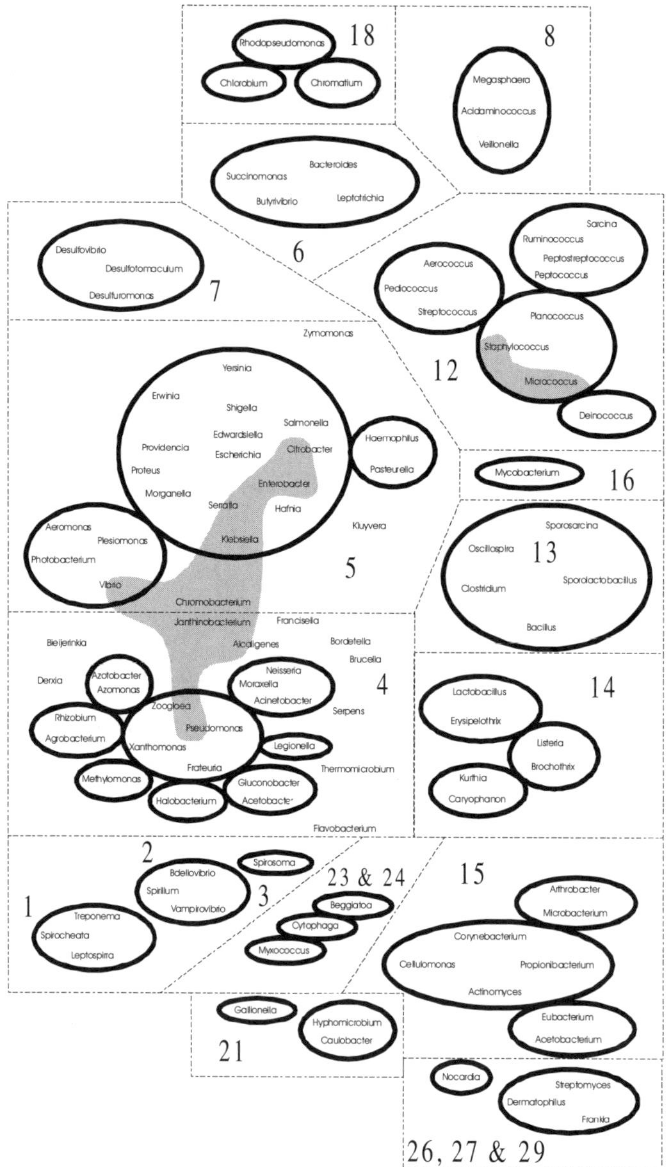

Figure 10.43 Common bacterial consortial (shaded) complexes in the slime-forming bacteria (SLYM) BART are dominated by the various bacteria that are able to generate copious slime (EPS) formations. In the DS - CL reaction pattern, the initial reaction is likely to be triggered by Section 5 enteric bacteria that can generate dense slimes (e.g., *Enterobacter* and *Klebsiella*) followed by a more broadly based consortium that causes the reaction to shift to a general cloudiness.

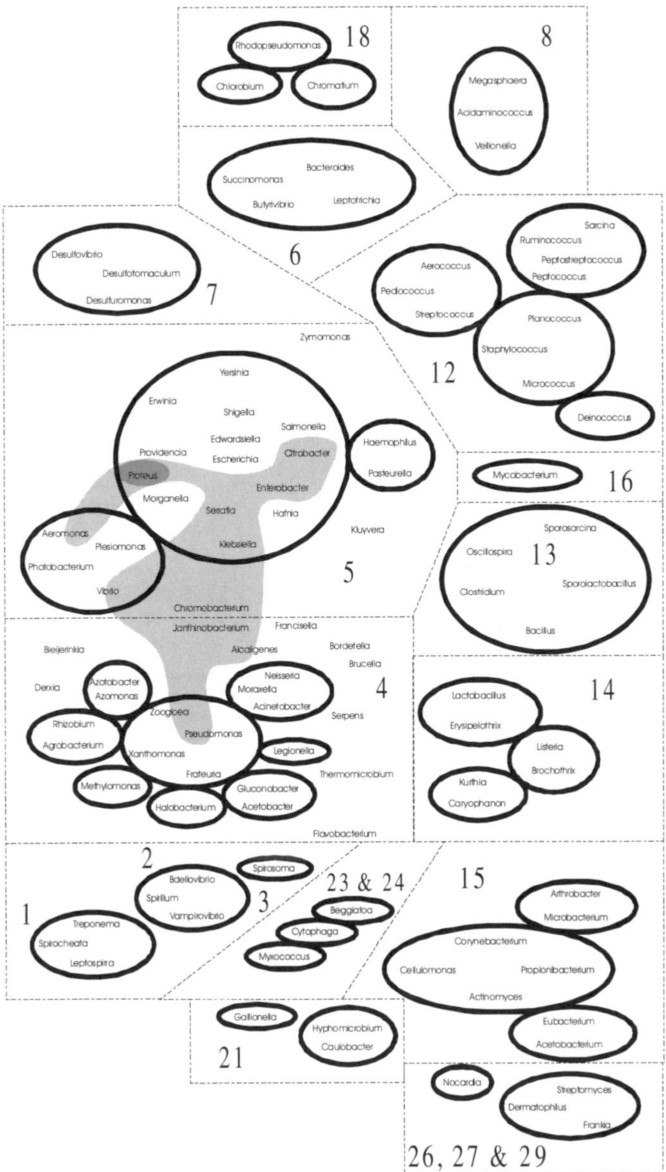

Figure 10.44 Common bacterial consortial (shaded) complexes in the slime-forming bacteria (SLYM) BART are dominated by the various bacteria that are able to generate slime (EPS) formations. In the CP - CL reaction pattern, the initial reaction is triggered by lateral (density controlled) plates that form usually at the redox front. This is commonly caused by *Proteus* and is followed by the growth of a more broadly based consortium that causes the reaction to shift to a general cloudiness.

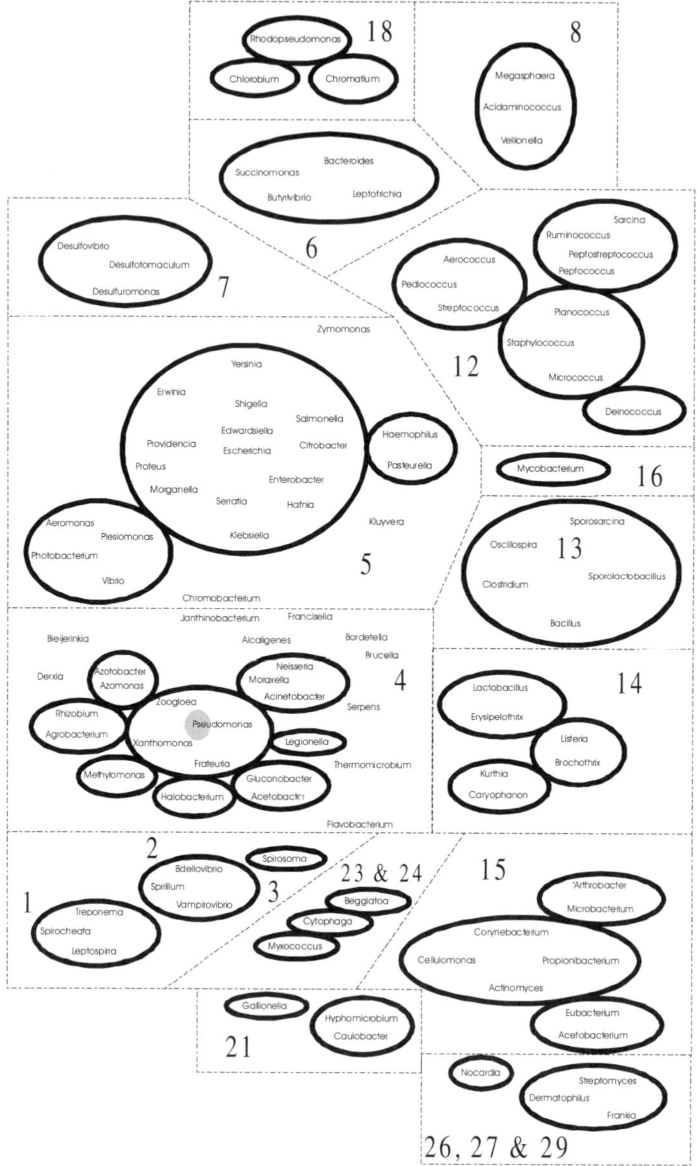

Figure 10.45 Common bacterial consortial (shaded) complexes in the slime-forming bacteria (SLYM) BART are dominated by the various bacteria that are able to generate slime (EPS) formations. In the CL - PB or the CL - GY reaction patterns, the dominant genus in the consortium (shown in atlas figure above) is *Pseudomonas* but only a limited range of species will cause the secondary PB or GY reactions.

APPENDIX

Listing of the BART™ Reaction Codes

The reactions codes listed below occur in the iron-related bacteria (IRB-BART™), the sulfate-reducing bacteria (SRB-BART™) and the slime-forming bacteria (SLYM-BART™) biodetectors. More information concerning these tests can be found in Chapter Five of the book "Microbiology of Well Biofouling" by D. Roy Cullimore published in 1999 by Lewis Publishers, Boca Raton. Each reaction code is described in sequence for each of the three biodetectors.

Reaction Codes for the IRB-BART™

CL, Clouded Growth (IRB-BART™)

When there are populations of aerobic bacteria, the initial growth may be at the REDOX front that commonly forms above the medium diffusion front. This growth usually takes the form of lateral or "puffy" clouding which is most often grey in color. It should be noted that if the observer tips the BART™ slightly, the clouds will move to maintain position within the tube. Commonly, the medium will be darker beneath the zone of clouding and lighter above.

BG, Brown Gel (IRB-BART™)

In this reaction, a basal, gel-like brown growth forms that maintains structure and position even when gently rotated or tilted. This brown gel can occupy the whole of the basal cone of the inner test vial and also extend up the sidewall of the inner test vial to a height of up to15 mm. The solution above the gel is commonly clear and colorless. Over time it is often noticed that the size of the gel mass will grow and later shrink. Detachment can happen so that a single brown gel-like mass can be seen floating in the test vial.

BC, Brown Cloudy (IRB-BART™)

Unless there is a very large population of IRB in the sample, this reaction is normally a secondary reaction (often following reactions CL, FO, or RC) and may be recognized as a dirty brown solution that may have a brown ring around the ball.

FO, Foam (IRB-BART™)

This is a very easy reaction to recognize since gas bubbles around the ball form a foam ring or sometimes the bubbles collect over greater than 50% of the underside of the ball. On some occasions, bubbles will collect on the walls of the inner test vial but this is not significant until the bubbles collect around the ball. The solution usually remains clear but commonly has a yellow or greenish-yellow color. The bubbles can at times be seen in the foam to be individually coated with slime that may give the bubbles a color ranging from brown through to orange, yellow or grey. Sometimes when integrated together into a foam, this foam is tough enough to either "lift" the FID ball out of the liquid solution or submerge the FID ball below the surface of the liquid solution. Do not confuse this reaction with the generation of bubbles (usually randomly) when oxygen supersaturates as the sample temperature comes up from a lower temperature (of the sample's source). These bubbles are recognized as being reflective and not bound in any slime and dispersed within the inner test vial under the ball and on the walls. They usually disappear within two days.

This FO reaction is most commonly related to a sample in which many microbes are functioning anaerobically. It can often be "harmonized" with the presence of SRB (reactions BB, BT or BA in the SRB-BART™). In other words, the occurrence of a FO in the IRB-BART™ can often be followed by a positive detection of SRB in the SRB-BART™ if that test has been performed on the same sample.

RC, Red, Slightly Clouded (IRB-BART™)

The liquid medium remains a clear to a dark reddish solution. The solution will cloud fairly quickly and shift to a BC reaction generally after a BR has formed around the ball.

BR, Brown Ring (IRB-BART™)

A reddish-brown to dark brown slime ring forms around the ball. This ring is entire and tight and usually around 3 mm in width. Generally, the brown slime ring will sit between the liquid surface and the equator of the ball and commonly intensifies over time. On some occasions, this reaction possesses an unusual feature in that the slime ring can "bio-lock" the ball to the walls of the test vial. In these cases, when the test vial is turned upside down, the ball remains (glued) in place and the liquid remains above the ball. What has happened is that the ring has become formed biologically into an hydraulic barrier.

GC, Green Clouded (IRB-BART™)

Solution goes to a shade of green and becomes cloudy without, necessarily, the formation of defined clouds or gel-like forms. No slime ring is formed around the FID. This cloudiness will gradually increase and often this reaction will shift to a dark green very cloudy solution. As the solution becomes a darker green and cloudier, a BR reaction may form but this is usually fairly thin.

BL, Blackened Liquid (IRB-BART™)

This is commonly a secondary or tertiary reaction rather than an initial reaction. It is recognized as a clear, often colorless, solution surrounded by large blackened zones in the basal cone of the inner test vial.

Note that if "fuzzy" growths form around the ball of the IRB-BART™, there are fungal spores present. This happens occasionally where a water sample has travelled through a semi-saturated zone. These create a reaction in which a white, grey or speckled "fuzzy" mat forms around and even over the ball. The upper surface of the mat often forms into a tight mass with an irregular surface. The lower surface of the mat can often be seen to be extending into the liquid medium by thread-like processes 2 to 5 mm in length. These growths may bio-lock the ball to the wall of the inner test vial for a period of time. Solution usually remains fairly clear but globular-like deposits may be present. Solution may cloud over time. This reaction is caused by the presence of large populations of fungal spores in the water.

Reaction Codes for the SRB-BART™

BB, Blackened Base (SRB-BART™)

The reaction is recognizable by the formation of a blackened deposit in the basal cone of the test vial. It may be first observed by looking up into the underside of the cone of the inner test vial. Blackening frequently starts as a 2- to 3-mm wide ring around the central peg and gradually spreads outwards. Black specking may also occur on the bottom 15 mm of the walls of the test vial immediately above the cone. The liquid medium should be clear (see reaction CG below) and there should be commonly no slime ring around the ball.

BT, Blackening around the Ball (SRB-BART™)

A slime ring may be viewed around the ball with patches of black specking or zones intertwined in the slime growths. The slime itself is not a characteristic of this reaction but the blackening is. The slime usually is either a white, grey, beige, or yellow color and tends to form on the sides of the ball. The blackening often begins as a speckling which gradually expands to patches within the slime.

BA, Combination of BB and BT (SRB-BART™)

A combination of reactions BB and BT constitute a reaction BA. Blackening occurs both in the base and around the ball although the length of the inner test vial may not be blackened.

CG, Cloudy Gel-Like (SRB-BART™)

While not a positive indication for the presence of SRB, this reaction is recognized since it does indicate the presence of anaerobic bacteria and often precedes the generation of reactions BB, BT or BA. It is recognized by the appearance of cloud-like structures in the colorless liquid medium. Usually these form from the bottom up and initially at a height of 20 to 25 mm up the side-wall of the inner test vial. This clouded zone may expand to render the liquid medium turbid. These clouds are relatively stable structures and have defined edges. This reaction does not mean that SRB are present but that anaerobic bacteria are. Commonly, the CG reaction precedes the blackening or occur shortly after the commencement of the blackening.

Reaction Codes for the SLYM-BART™

DS, Dense Slime (SLYM-BART™)

This reaction may not be obvious and require the observer to gently rotate the BART™ test at which time slimy deposits swirl up. These deposits may swirl in the form of a twisting slime when the tube is gently rotated. This swirl can reach 40 mm up into the liquid column, or it may rise up as globular gel-like masses that settle fairly quickly. Once the swirl has settled down, the liquid may become clear again. In the latter case, care should be taken to confirm that the artifact is biological (ill-defined edge, mucoid, globular) rather than chemical (defined edge, crystalline, often white or translucent). Generally, these dense slime growths are beige, white or yellowish-orange in color.

CP, Cloudy Plates Layering (SLYM-BART™)

When there are populations of aerobic bacteria, the initial growth may be at the REDOX front that commonly forms above the yellowish-brown diffusion front. This growth usually takes the form of lateral or "puffy" clouding which is most commonly grey in color. Often the lateral clouds may be disk-like in shape (plates) and relatively thin (1 to 2 mm). It should be noted that if the observer tips the BART™ slightly, the clouds or plates often move to maintain position within the tube. The edges of the plates are distinct while the edges of the "puffy" forms of layering are indistinct. These formations are most commonly observed 15 to 30 mm beneath the fill line, while cloud formations will tend to extend to cause an overall cloudiness of the liquid medium. These plates sometimes appear to divide (multiple plating) before coalescing into a cloudy liquid medium.

SR, Slime Ring (SLYM-BART™)

A slime ring, usually 2 to 5 mm in width, forms on the sides of the ball. The appearance is commonly mucoid and may be a white, beige, yellow, orange or violet color that commonly becomes more intense over time on the upper edge.

CL, Cloudy Growth (SLYM-BART™)

Solution is very cloudy and there may sometimes be a poorly defined slime growth around the ball. Sometimes a glowing may be noticed in at least a part of the top 18mm of the liquid medium. This glowing is due to the generation of U.V. fluorescent pigments commonly by some species of *Pseudomonas*. The common pigments doing this are a pale blue (PB) or a yellowish green (YG) color. Note that this glowing may not be readily observable unless a U.V. light is used. The occurrence of the glowing in a U.V. light means that there is a probability of potentially pathogenic species of *Pseudomonas* and confirmatory testing is recommended.

BL, Blackened Liquid (SLYM-BART™)

This is commonly a secondary or tertiary reaction rather than an initial reaction. It is recognized as a clear, often colorless, solution that is surrounded by large blackened zones in the basal cone and up the walls of the test vial. The BL often parallels the BL reaction in the IRB when the two BART™ types are used together on the same sample.

TH, Thread-Like Strands (SLYM-BART™)

On some occasions, the slime forms into threads that form web-like patterns in the liquid medium. Sometimes, these threads interconnect from the ball to the floor of the inner test vial.

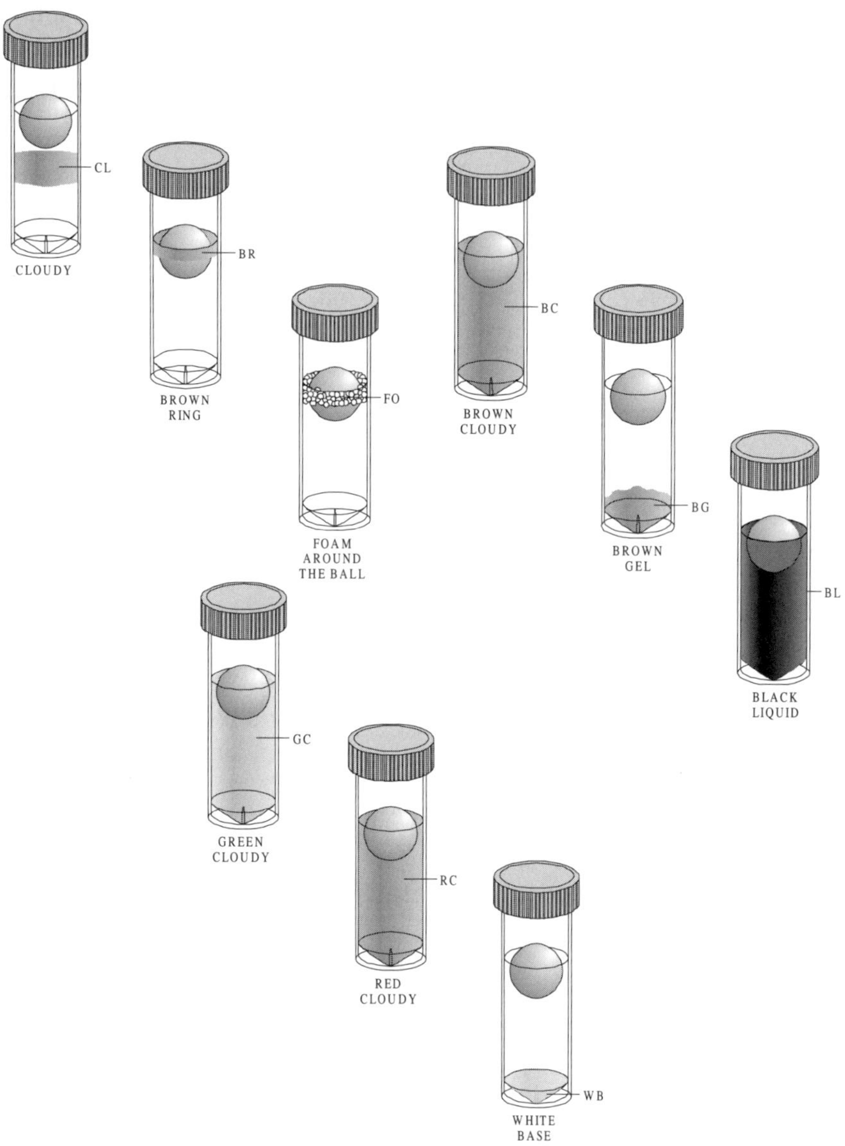

Figure A.1 Diagrammatic presentation of the nine major reaction types seen in an iron-related bacteria (IRB) BART when incubated at room temperature using either 15ml of a water sample or 0.1g of a solid or semi-solid material. The reactions usually occur in a sequence within ten days when these bacteria are present. Note that WB is not an interpretable reaction.

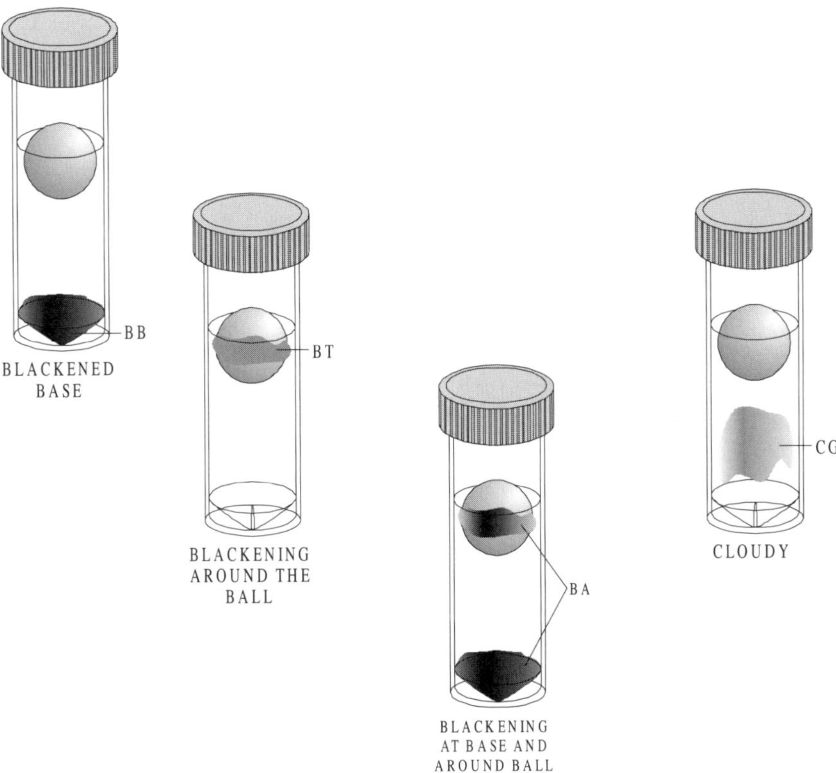

Figure A.2 Diagrammatic presentation of the nine major reaction types seen in a sulfate-reducing bacteria (SRB) BART when incubated at room temperature using either 15ml of a water sample or 0.1g of a solid or semi-solid material. The reactions usually occur in a sequence within ten days when these bacteria are present.

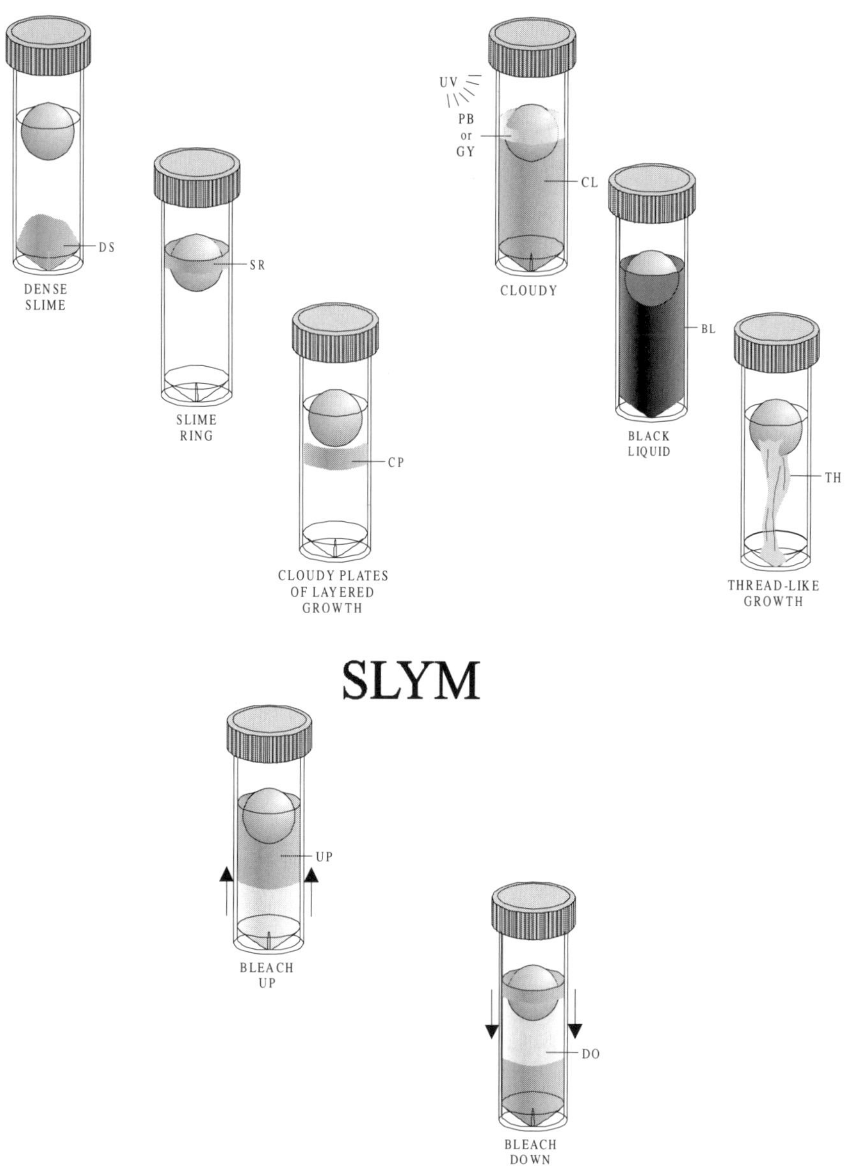

Figure A.3 Diagrammatic presentation of the nine major reaction types seen in a slime-forming bacteria (SLYM) BART (upper reaction set) and a heterotrophic aerobic bacteria (HAB) BART (lower reaction set) when incubated at room temperature using either 15ml of a water sample or 0.1g of a solid or semi-solid material. The reactions usually occur in a sequence within ten days when these bacteria are present.

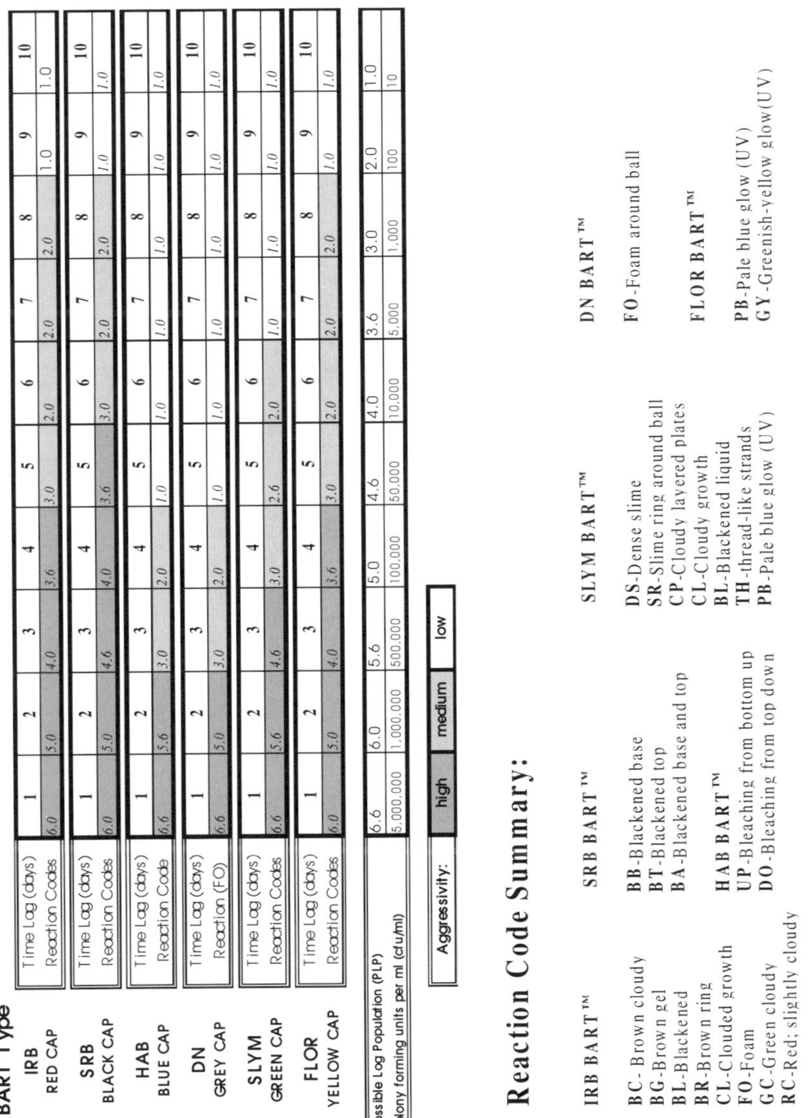

BART Type		1	2	3	4	5	6	7	8	9	10
IRB RED CAP	Time Lag (days)	1	2	3	4	5	6	7	8	9	10
	Reaction Codes	6.0	5.0	4.0	3.6	3.0	2.0	2.0	2.0	1.0	1.0
SRB BLACK CAP	Time Lag (days)	1	2	3	4	5	6	7	8	9	10
	Reaction Codes	6.0	5.0	4.6	4.0	3.6	3.0	2.0	2.0	1.0	1.0
HAB BLUE CAP	Time Lag (days)	1	2	3	4	5	6	7	8	9	10
	Reaction Code	6.6	5.6	3.0	2.0	1.0	1.0	1.0	1.0	1.0	1.0
DN GREY CAP	Time Lag (days)	1	2	3	4	5	6	7	8	9	10
	Reaction (FO)	6.6	5.0	3.0	2.0	1.0	1.0	1.0	1.0	1.0	1.0
SLYM GREEN CAP	Time Lag (days)	1	2	3	4	5	6	7	8	9	10
	Reaction Codes	6.6	5.6	4.6	3.0	2.6	2.0	1.0	1.0	1.0	1.0
FLOR YELLOW CAP	Time Lag (days)	1	2	3	4	5	6	7	8	9	10
	Reaction Codes	6.0	5.0	4.0	3.6	3.0	2.0	2.0	2.0	1.0	1.0
Possible Log Population (PLP)		6.6	6.0	5.6	5.0	4.6	4.0	3.6	3.0	2.0	1.0
colony forming units per ml (cfu/ml)		5,000,000	1,000,000	500,000	100,000	50,000	10,000	5,000	1,000	100	10

Aggressivity:	high	medium	low

Reaction Code Summary:

IRB BART™

BC- Brown cloudy
BG-Brown gel
BL-Blackened
BR-Brown ring
CL-Clouded growth
FO-Foam
GC-Green cloudy
RC-Red; slightly cloudy

SRB BART™

BB-Blackened base
BT-Blackened top
BA-Blackened base and top

HAB BART™

UP-Bleaching from bottom up
DO-Bleaching from top down

SLYM BART™

DS-Dense slime
SR-Slime ring around ball
CP-Cloudy layered plates
CL-Cloudy growth
BL-Blackened liquid
TH-thread-like strands
PB-Pale blue glow (UV)

DN BART™

FO-Foam around ball

FLOR BART™

PB-Pale blue glow (UV)
GY-Greenish-yellow glow(UV)

Figure A.4 Interpretation of BART reactions and time lags (to the first observation of a positive reaction) can be recorded on the BART data interpretation sheet which allows the first observation time (time lag) to be used to project the log cfu/ml population in the sample under test. The reaction patterns can be entered as they occur and this forms a reaction pattern signature which allows the form of the bacterial consortium to be determined.

Selected Bibliography

Balows, A., Trüper, H.G., Dworkin, M., Harder, W., and Schleifer, K.H. (1991) *The Prokaryotes*, 2nd ed., Springer-Verlag, New York.

Barrow, G.I. and Feltham, R.K.A., Eds. (1993) *Cowan and Steel's Manual for the Identification of Medical Bacteria,* 3rd ed., Cambridge University Press, Cambridge.

Board, R.G., Jones, D., and Skinner, F.E., Eds. (1992) *Identification Methods in Applied and Environmental Microbiology,* Blackwell Scientific, Oxford.

Collins, C.H. and Grainge, J.M., Eds. (1985) *Isolation and Identification of Microorganisms of Medical and Veterinary Significance*, Academic Press, London.

Cullimore, D.R. (1992) *Practical Manual of Groundwater Microbiology*, Lewis Publishers, Boca Raton, FL.

Geldrich, E.E. (1996) *Microbial Quality of Water Supply in Distribution Systems*, Lewis Publishers, Boca Raton, FL.

Goodfellow, M. and Minnekin, D.E., Eds. (1993) *Handbook of New Bacterial Systematics,* Academic Press, London.

Holt, J.G., Krieg, N.R., Sneath, P.H.A., Staley, J.T. and Williams, S.T. (1984) Vol. 4, *Actinomycetes* in *Bergey's Manual of Systematic Bacteriology*, Williams & Wilkins, Baltimore.

Holt, J.G., Krieg, N.R., Sneath, P.H.A., Staley, J.T. and Williams, S.T. (1984) Vol. 1, *Spirochaetes, Gram-negative bacteria, rickettsiae, chlamydiiae and mollicutes* in *Bergey's Manual of Systematic Bacteriology*, Williams & Wilkins, Baltimore.

Holt, J.G., Krieg, N.R., Sneath, P.H.A., Staley, J.T. and Williams, S.T. (1989) Vol. 3, *Autotrophs, budding bacteria, gliding bacteria, archaebacteria* in *Bergey's Manual of Systematic Bacteriology*, Williams & Wilkins, Baltimore.

Leboffe, M.J. and Pierce, B.E. (1996) *A Photographic Atlas for the Microbiology Laboratory*, Morton, Englewood, Colorado.

Logan, N.A. (1994) *Bacterial Systematics*, Blackwell Scientific, London.

Park, W.H. and Williams, A.W. (1910) *Pathogenic Micro-Organisms including Bacteria and Protozoa, A Practical Manual for Students, Physicians and Health Officers*, Lea & Febiger, Philadelphia.

Prévot A. and Fredette, V. (1966) *Manual for the Classification and Determination of the Anaerobic Bacteria* (1st American ed.), Lea & Febiger, Philadelphia.

Skerman, V.B.D. (1967) *A Guide to the Identification of the Genera of Bacteria* (2nd ed.), Williams & Wilkins, Baltimore.

Stackebrandt, E. and Goodfellow, M., Eds. (1991) *Nucleic Acid Techniques in Bacterial Systematics,* John Wiley & Sons, Chichester.

Weisberg, W.G., Barns, S.M., Pelletier, D.A. and Lane, D.J. (1991) 16S ribosomal DNA amplification for phylogenetic study, *J. of Bacteriol.,* 173.

Index

A

B

D

E

F

G

H

N

O

T

U

V